I0787954

Advanced CMOS-Compatible Semiconductor Devices 19

Editors:

J. Martino

B.-Y. Nguyen

F. J. Gamiz

H. Ishii

J.-P. Raskin

S. Selberherr

E. Simoen

Sponsoring Divisions:

 Electronics and Photonics

Dielectric Science and Technology

Published by
The Electrochemical Society
65 South Main Street, Building D
Pennington, NJ 08534-2839, USA

tel 609 737 1902
fax 609 737 2743
www.electrochem.org

ecstransactions ™

Vol. 97, No. 5

Copyright 2020 by The Electrochemical Society.
All rights reserved.

This book has been registered with Copyright Clearance Center.
For further information, please contact the Copyright Clearance Center,
Salem, Massachusetts.

Published by:

The Electrochemical Society
65 South Main Street
Pennington, New Jersey 08534-2839, USA

Telephone 609.737.1902
Fax 609.737.2743
e-mail: ecs@electrochem.org
Web: www.electrochem.org

ISSN 1938-6737 (online)
ISSN 1938-5862 (print)
ISSN 2151-2051 (cd-rom)

ISBN 978-1-62332-604-3 (CD-ROM)
ISBN 978-1-62332-605-0 (USB)
ISBN 978-1-60768-893-8 (PDF)

Printed in the United States of America.

Preface

The papers included in this issue of *ECS Transactions* were approved for the symposium "Advanced CMOS-Compatible Semiconductor Devices 19" of the 237th Meeting of The Electrochemical Society in Montreal, CA, from May 10-15, 2020.

This historical symposium started more than 30 years ago, focused on SOI-materials, technology, and devices. It was renamed in 2011 in order to cover recent significant advances not only in SOI-based nanoelectronics technologies and devices, but also in advanced CMOS compatible devices, circuits, and applications using bulk MOSFETs, fully depleted devices, multi-gate devices (double gate, FinFET, triple gate, nanowire, nanosheet), junctionless FET, high-power devices, tunnel-FETs, semiconductor sensors, memory device, and spintronics.

This issue of *ECS Transactions* contains 21 papers, including three high-level keynotes and five invited papers with the participation of many worldwide research groups from North America, Europe, Asia, and South America. Many of the papers include collaborations between researchers from institutions of different continents. This issue of *ECS Transactions* has been organized in four sections: Devices Physics and Characterization, Sensors and Biosensors, Process Technology, and Memory and Circuits. I would like to thank all the authors for submitting excellent manuscripts on time.

It is my great pleasure to acknowledge the symposium co-organizers Francisco Gamiz, Hiromu Ishii, Bich-Yen Nguyen, Jean-Pierre Raskin, Siegfried Selberherr, and Eddy Simoen on their active involvement and efforts in spite of their heavy workload. Many thanks to ECS staff members for their support and assistance.

On behalf of the organizers, I would like to express our great gratitude for the keynotes and invited speakers who set the quality for the meeting, and I thank all symposium speakers and coauthors for contributing with this symposium. The organizers thank the ECS Electronics and Photonics Division for its financial and technical support.

Joao Antonio Martino
University of Sao Paulo, Brazil
May 2020

ECS Transactions, Volume 97, Issue 5
Advanced CMOS-Compatible Semiconductor Devices 19

Table of Contents

Preface   *iii*

Chapter 1
Devices, Physics, and Characterization

*(Keynote)*Â Gate-All-Around Nanowire & Nanosheet FETs for Advanced, Ultra-   3
Scaled Technologies
   A. Veloso, P. Matagne, D. Jang, T. Huynh-Bao, A. Chasin, E. Simoen, G. Eneman,
   A. De Keersgieter, H. Mertens, N. Horiguchi

*(Keynote)*Â Fully-Depleted SOI TechnologyÂ for High-Power RF Applications   15
   T. V. Dinh

*(Invited)*Â Advanced Transistors for High Frequency Applications   27
   B. Parvais, U. Peralagu, A. Vais, A. Alian, L. Witters, Y. Mols, A. Walke, M. Ingels,
   H. Yu, V. Putcha, A. Khaled, R. Rodriguez, A. Sibaja-Hernandez, S. Yadav,
   R. ElKashlan, M. Baryshnikova, G. Mannaert, R. Alcotte, B. Kunert, E. Simoen,
   E. Zhao, B. De Jaeger, D. Fleetwood, R. Langer, M. Zhao, P. Wambacq,
   N. Waldron, N. Collaert

*(Invited)*Â Junctionless Device Cross-Section: A Key Aspect for Overcoming   39
Boltzmann Tyranny
   A. Kranti, M. Gupta

Discussion on the Figures of Merit of Identified Traps Located in the Si Film: Surface   45
Versus Volume Trap Densities
   B. Cretu, B. Nafaa, E. Simoen, G. Hellings, D. Linten, C. Claeys

Impact of Gate Dielectric Material on Basic Parameters of MO(I)SHEMT Devices   53
   P. G. D. Agopian, G. J. D. Carmo, J. A. Martino, E. Simoen, N. Collaert

On the Correlation Between Static and Low-Frequency Noise Parameters of Vertical
Nanowire Â nMOSFETs   59
E. Simoen, A. Chasin, P. Matagne, E. Rosseel, A. Y. Hikavyy, R. Loo, P. Favia,
H. Bender, E. Vancoille, A. Veloso

Intrinsic Voltage Gain of Stacked GAA Nanosheet MOSFETs Operating at High   65
Temperatures
W. F. Perina, V. C. P. Silva, J. A. Martino, P. G. D. Agopian, E. Simoen, A. Veloso

Electrical Behavior of Effects LCE and PAMDLE of the Ellipsoidal MOSFETs in a   71
Huge Range of High Temperatures
E. H. S. Galembeck, S. P. Gimenez

Chapter 2
Sensor and Biosensor

(Invited)Â Post-CMOS Compatible Silicon MEMS Nano-Tactile Sensor for Touch   79
Feeling Discrimination of Materials
H. Takao

(Invited)Â CMOS-MEMS Based Microgravity Sensor and Its Application   91
K. Masu, K. Machida, D. Yamane, H. Ito, N. Ishihara, T. F. M. Chang, M. Sone,
R. Shigeyama, T. Ogata, Y. Miyake

Influence of the Biomaterial Thickness in a Dielectrically Charged Modulated   109
Fringing Field Bio-Tunnel-FET Device
C. N. Macambira, P. G. D. Agopian, J. A. Martino

Study of a Charge-Based Biosensor and Reconfigurability using BESOI MOSFET   115
L. S. Yojo, R. C. Rangel, K. R. A. Sasaki, J. A. Martino

Study of Underlapped Finfets Behavior for a Radiation Sensing Purpose   121
W. D. S. Fonseca, P. G. D. Agopian

Chapter 3
Process Technology

(Keynote)Â Low Temperature SmartCut™ Process for 3D Integration 129
  W. Schwarzenbach, S. Reboh, A. Ghorbel, G. Gaudin, G. Besnard, F. Mazen,
  V. Loup, S. Maitrejean, L. Brunet, B. Y. Nguyen, C. Maleville

(Invited)Â Dielectric Science on Today's Devices 135
  D. Misra

Gate Electrode Material Effect on Characteristics of Zirconium-Doped Hafnium 143
Oxide High-k MOS Capacitors
  W. S. Lin, L. Liu, Y. Kuo

Graphitic Nanoporous Carbon Thin Films: Fabrication Method, Structural, Electrical 151
and Gas Sensor Properties
  O. M. Slobodian, Y. V. Gomeniuk, A. V. Vasin, A. V. Rusavsky, P. N. Okholin,
  O. Y. Gudymenko, O. Y. Khyzhun, A. Nikolenko, P. Lytvyn, A. Korchovyy,
  R. Yatskiv, T. M. Nazarova, V. Stepanov, D. Kisyl, A. N. Nazarov

Chapter 4
Memory and Circuits

Influence of Current Redistribution in Switching Models for Perpendicular STT- 159
MRAM
  S. Fiorentini, R. Lacerda de Orio, S. Selberherr, J. Ender, W. Goes, V. Sverdlov

Readout Circuit Design Using Experimental Data of Line-TFET Devices 165
  W. GonÂ§alez, R. Rangel, J. A. Martino, P. G. D. Agopian

Proton-Irradiation Influence on Current Mirror Circuit Using Verilog-A Approach 171
Based on Experimental SOI FinFET Characteristics
  B. R. de Sousa, P. G. D. Agopian, J. A. Martino

Author Index 179

Facts about ECS

The Electrochemical Society (ECS) is an international, nonprofit, scientific, educational organization advancing the theory and practice of electrochemistry and solid state science and technology, and allied subjects. The Society was founded in Philadelphia in 1902 and incorporated in 1930. There are currently over 8,400 members from around the globe representing 13 technical division and 23 geographical sections and a growing student membership program with almost 70 student chapters. The Society is also supported by more than 800 corporations, government agencies and academic institutions through institutional membership, corporate programs and subscriptions.

The technical activities of the Society are carried on by divisions. Sections of the Society host symposia, programs and events focused on their respective geographic regions. Major international meetings of the Society are held in the spring and fall of each year. At these meetings, the divisions and partnered organizations hold general sessions and sponsor symposia on specialized subjects.

The Society has an active publication program that includes the following:

Journal of The Electrochemical Society — (JES) is the flagship journal of The Electrochemical Society and the oldest peer-reviewed journal in its field. Since its founding in 1902, JES has evolved into one of the most highly cited and prestigious journals in electrochemistry and materials science with a cited half-life of greater than 10 years.

ECS Journal of Solid State Science and Technology — (JSS) is one of The Electrochemical Society's newest peer-reviewed journals. Launched in 2012, JSS covers fundamental and applied areas of solid state science and technology, including experimental and theoretical aspects of the chemistry and physics of materials and devices.

ECS Transactions — (ECST) is an online database of technical papers—a high-quality venue for authors and an excellent resource for researchers. ECST offers the full-text content of proceedings from ECS meetings and ECS sponsored conferences.

The Electrochemical Society Interface — *Interface* is a quarterly news magazine that provides a forum for the lively exchange of ideas within the scientific community. Published online with free access in the ECS Digital Library, issues engage readers with a featured scientific research section, and summaries of research breakthroughs, and industry and Society news.

ECS Meeting Abstracts — Published 2-4 times a year, this publication contains extended abstracts of technical papers presented at ECS biannual meetings and ECS-sponsored meetings and offers a first look into the current research in the field.

ECS Monograph Series — ECS monographs provide authoritative, detailed accounts of specific topics in electrochemistry and solid state science and technology. These titles are sponsored by ECS and published in cooperation with noted publishers such as John A. Wiley & Sons.

For more information on these and other Society activities, visit the ECS website:

www.electrochem.org

CHAPTER 1

Devices, Physics, and Characterization

Gate-All-Around Nanowire & Nanosheet FETs for Advanced, Ultra-Scaled Technologies

A. Veloso, P. Matagne, D. Jang, T. Huynh-Bao, A. Chasin, E. Simoen, G. Eneman, A. De Keersgieter, H. Mertens, and N. Horiguchi

Imec, Kapeldreef 75, 3001 Leuven, Belgium

We report on gate-all-around (GAA) vertically stacked lateral nanowires (NW) and nanosheets (NS) FET devices as promising candidates to replace finFETs and help preserve the power, performance, area, and cost (PPAC) logic roadmap for advanced sub-5nm technology nodes. In addition, GAA vertical NW/NS FETs appear particularly attractive for enabling highly dense memory cells such as ultra-scaled MRAMs with lower energy consumption and smaller access latency values. Key fabrication challenges and device features for both types of transistors will be discussed here, together with the possibility to manufacture them on the same wafer by a cost-effective, co-integration scheme to obtain simultaneously a high-performance logic platform and have increased on-chip memory content.

Introduction

As two-dimensional (2D) CMOS scaling is reaching its physical limits, maintaining profitable node-to-node cost-per-transistor gains requires increasingly constraining design restrictions. To help preserve the power, performance, area, and cost (PPAC) logic roadmap, several innovations are being explored, such as cell height reduction by decreasing the number of metal tracks (1-4). This allows compensating for the recent trend of a more modest pitch scaling, but it also requires fin depopulation from two fins to one fin per finFET transistor for ultra-scaled cells with reduced number of metal tracks (3,4) which impacts the device performance. As a result, in order to keep the scaling gains, introduction of novel device architectures is needed, with gate-all-around (GAA) lateral nanowire (NW) or nanosheet (NS) FETs (4-31) being regarded as the most promising and mature finFET replacements for advanced sub-5nm technology nodes. These devices provide better electrostatics, allowing further gate length (L_{gate}) scaling, with NS offering higher current drivability (I_{ON}) per layout footprint than NW thanks to larger effective widths (W_{eff}). At the same time, they share many of the building blocks of a finFET integration flow which enables implementation of a smoother technological transition with regard to manufacturing as compared to other transistor architectures.

Another type of GAA devices, with vertical NW (VNW) or vertical NS (VNS) and vertically defined L_{gate}, has been shown to have the potential to enable denser and more energy efficient SRAMs and MRAMs when used as the cell transistors or selector, respectively (14,32-37). In the overall scaling roadmap, the latter type of memory has been gaining increased momentum as an alternative for enabling large, ultra-high density, last-level caches for systems with reduced area and energy (38-42). On the whole, GAA VNW/VNS FETs represent a more disruptive technological transition, moving from 2D to 3D layouts. Nevertheless, they can also open up new scaling paths in the third dimension. With this in mind, and in view of the overall rising system demands which include an increased need for more on-chip memory content, it is worthwhile to consider a cost-effective way to fabricate both types of GAA devices on the same wafer, targeting high-performance logic or for specialized memories, as will be addressed in this paper.

Lateral Nanowire and Nanosheet FETs

NS FETs have an advantage over NW FETs in terms of enabling higher current drivability per layout footprint on the wafer thanks to their larger effective widths. Their expected larger I_{ON} values have also been confirmed by technology computer-aided design (TCAD) simulations (43) due to an enhanced volume inversion in the on-state and to a better subband occupation which leads to higher injection velocity (4). Furthermore, by having the option to vary their W_{eff} on the same chip/wafer, these devices enable additional design flexibility, typically considering the use of wider NS for targeting high-performance computing and of narrower NS for low-power applications (18,21,26).

Fig. 1 shows a benchmark of the power-performance values calculated for inverter based ring oscillators (RO). The results correspond to a fan-out of three (FO3) with ~50× contacted-gate-pitch (CGP) length for back-end-of-line (BEOL) load, assuming representative design rules for various technology nodes: from 7nm (N7) down to 2nm (N2) nodes. The following assumptions for CGP / metal-pitch (MP) were used: 56/40 → 48/28 → 45/21 → 42/16 (dimensions in nanometers) for N7 → N5 → N3 → N2, respectively. The data show that, while finFET technology is still projected to deliver gains when scaling down from the 7nm to the 5nm nodes using two fins-per-device based cells, additional performance gains at iso-V_{DD} are compromised beyond that. Indeed, assuming that at the 3nm node the cell height is shrunk to five metal tracks, a change into a one fin-per-device scenario for finFET based circuits is required to enable such scaled cells. Furthermore, incorporation of a buried metal line (BML) or buried power rail

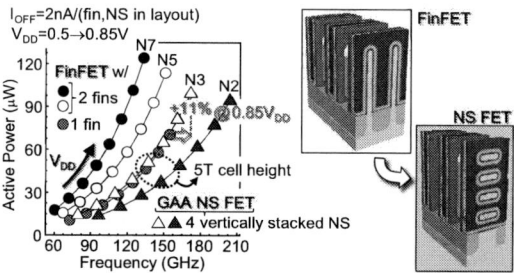

Figure 1. Simulated power-performance's benchmark for inverter ring oscillators (RO) with fan-out of three and ~50× contacted-gate-pitch (CGP) length for back-end-of-line load. FinFET based cells with two fins-per-device for 7nm (N7) and 5nm (N5) nodes are compared with cells representative of further scaled nodes: 3nm (N3) and 2nm (N2). The last two assume the use of buried power rail technology and: a) finFETs with one fin-per-device for N3, or b) GAA NS FETs consisting of four vertically stacked NS, each with a width of 16nm (N3) or 13nm (N2). Overall, these data show that NS FETs are predicted to outperform finFETs at N3.

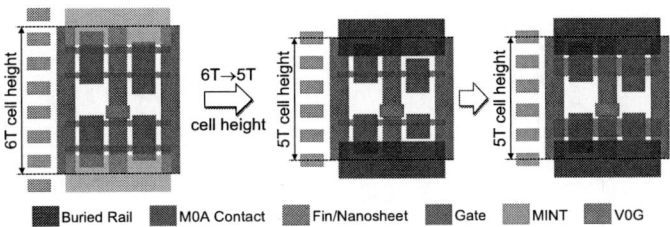

Figure 2. Illustration of standard cell layouts for N5 and N3 corresponding to 6 or 5 tracks (T) cell height, respectively. FinFET or NS FET devices are used as the cell transistors as described in Fig. 1, with introduction of a buried metal line or buried power rail technology also projected to occur at N3.

technology (44,45) is also assumed in our calculations to be implemented at this node as is illustrated in Fig. 2. In this context, it becomes therefore advantageous to introduce NS FETs at this point as they can outperform finFETs by delivering faster ROs at a given V_{DD} or by enabling power savings at matched performance: 15%, 19% frequency gains at V_{DD}=0.7V are obtained with NS FETs in Fig. 1 for the 3nm and 2nm nodes, respectively.

A strategy to further boost I_{ON} for NS FETs is to vertically stack several NS per transistor, increasing their total W_{eff}. This is an approach that can be done without impacting the devices' footprint on the wafer. It is validated in Fig. 3a wherein faster ROs, at similar V_{DD}, are obtained by increasing the number of vertically stacked NS per device, though at the expense of exhibiting higher power consumption values. The latter is a result from the increase in capacitance occurring for a multiple-sheets structure, which gets higher as more NS are stacked on top of each other. Minimizing the overall capacitance of these devices is therefore crucial for delivering both optimal performance and power efficiency.

To address this issue, introduction of a new process module in the fabrication flow of these devices is a required element for their successful implementation: the build-up of inner spacers in-between the stacked NS (4,16,20). Their integration can be done just prior to the source/drain (S/D) epitaxial growth, by performing first a controlled lateral SiGe etch (with high selectivity towards Si) in the Si/SiGe multi-layer fins, wherein the SiGe films play the role of sacrificial layers to form the Si channels. This is then followed by the filling of the created cavities with the so-called inner spacers dielectric. As discussed in Ref. (16), the first step may induce strain changes but epi growth in recessed S/D areas has been confirmed to still be effective for introducing a significant amount of stress in the vertically stacked Si channels for mobility boost.

Another knob to reduce the device parasitics consists in decreasing the vertical separation between the vertically stacked NS. This is demonstrated by the RO's frequency response improvement plotted in Fig. 3b with shrinking NS vertical pitch. An example of such hardware implementation is presented in Fig. 3c, for which the use of scaled Si/SiGe multi-layers growth prior to active level (fin) patterning is required. Indeed, since the Si NS are obtained by selectively removing the SiGe from the Si/SiGe multi-layer fins after dummy gate removal at the replacement metal gate (RMG) module (4,17,20,25,28,30), having thinner sacrificial SiGe layers leads to shorter vertical distances between the various Si nanosheets. At the same time, adopting this scaled device configuration also introduces additional challenges for NS FET fabrication, such as the need for thinner gate stacks that can fit in-between the nanosheets and that can

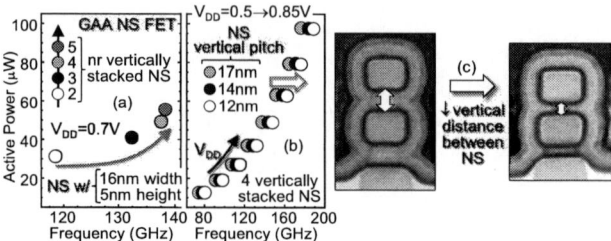

Figure 3. The speed of ring oscillators can be improved by: a) increasing the number of vertically stacked NS per device, with the drawback of having higher power consumption values; b) reducing the vertical distance between the stacked NS. Device implementation of the latter is illustrated by the annular bright-field scanning transmission electron microscopy (ABF-STEM) images in (c).

guarantee to provide both a similar gate control on all NS surfaces and to meet all the required gate stack electrical specs, e.g., in terms of EOT, leakage, V_T, reliability, variability, noise, etc. Achieving this poses some key technological challenges.

Fig. 4 illustrates one of the possible negative effects faced while trying to thin down the reference effective work-function (EWF) metal layer in the gate stack. Here, the reference n-type stack consists of a 4nm (nominal thickness) Al based metal layer m* (Figs. 4b,e). When thinning it down by > 50% (Figs. 4c,e), while the total physical thickness of the EWF metal stack is substantially and uniformly reduced around all the NS surfaces ($\sim$7.5nm $\rightarrow$ $\sim$4.2nm), that occurs at the expense of an increase in V_T ($V_{Tlin}\sim$0.24V $\rightarrow$ 0.6V). Upon exploration of various material options, recent promising results with an alternative n-type EWF metal n* have been disclosed in Ref. (28). Figs. 4d,f confirm that this layer n* can be deposited as a thinner film (<2.5nm thick), while Fig. 4a shows it can yield tighter V_T distributions and lower V_T values ($V_{Tlin}\sim$0.14V) as compared to those obtained for the thicker Al based reference stack. Additionally, no significant impact in gate leakage (I_G) and similar peak g_m values were reported in Ref. (28). Overall, the results demonstrate the feasibility and the successful use of a thinner gate stack in GAA NS FETs, without compromising their DC characteristics. Furthermore, this is achieved while simultaneously exhibiting improved

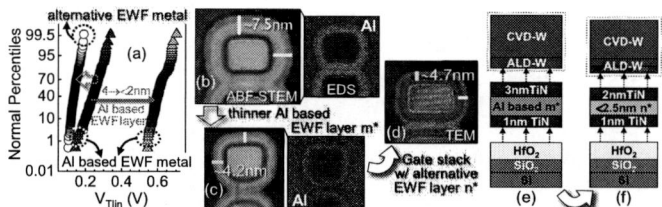

Figure 4. a) shows that increased V_{Tlin} values are obtained for n-type GAA NS FETs when thinning down the reference Al based EWF metal layer m* from 4nm (b) to <2nm (c) nominal thickness, while $\sim$100mV lower V_{Tlin} is achieved with an alternative EWF metal layer n*, thinner than 2.5nm (d). Schematics of the implemented gate stacks for the two types of EWF metals are depicted in (e) and (f).

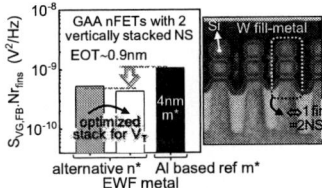

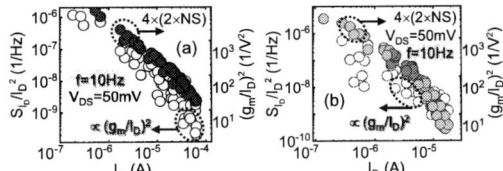

Figure 5. Slightly improved LF noise values are measured for GAA NS FETs consisting of two vertically stacked nanosheets and built using a thinner, alternative n* EWF metal vs. reference devices with 4nm Al based EWF metal layer m* in the gate stack.

Figure 6. SI_D/I_D^2 is proportional to $(g_m/I_D)^2$, at lower I_D, for devices built with different EWF metals in the gate stack, using: (a) a 4nm Al based m* reference, or (b) an alternative, thinner n* EWF metal layer. This is indicative of a similar dominant mechanism for the LF noise behavior exhibited by the two types of NS FETs.

noise and reliability characteristics.

Fig. 5 shows improved low-frequency (LF) noise behavior for n-type GAA NS FETs built with the alternative n* EWF metal layer as compared to the reference devices, with lower normalized input-referred noise spectral density (S_{VG}) values indicating less traps/defects present. Both types of devices exhibit 1/f noise behavior and, as shown in Fig. 6, proportionality of the normalized current noise spectral density (SI_D/I_D^2) vs. $(g_m/I_D)^2$, at lower I_D. The latter suggests that carrier number fluctuations are the dominant mechanism for the LF noise exhibited by the various NS FETs.

The EWF metal selection has been shown to be able to impact the profile of the oxide trap density (N_{ot}, derived from $S_{VG}.f$ (46)) vs. the trap depth (47). That is visible in Fig. 7a where a considerably uniform profile for devices with an Al based EWF metal layer (10) vs. a sloped profile for the case of a TiN EWF metal (10) are shown, despite similar N_{ot} values computed at the Si/SiO$_2$ interface. Similar behavior is also observed in vertically stacked NS FETs (Al based EWF metal shown in Fig. 7b), with devices built with the thinner n* EWF metal layer (in Fig. 7c) also exhibiting uniform N_{ot} profiles.

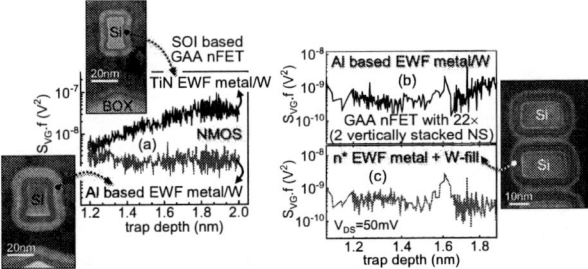

Figure 7. As shown in (a), the EWF metal choice can have an impact on the N_{ot} profiles (derived from $S_{VG}.f$) vs. trap depth in GAA NW FETs: sloped (and with N_{ot} increasing towards the metal gate) in case of TiN based devices, or uniform for the Al based gate stacks described in Ref. (10). Uniform N_{ot} vs. trap depth profiles are also extracted for GAA FETs with two vertically stacked nanosheets and built using either the (b) Al based reference or the (c) thinner EWF metal layers from Figs. 4-6.

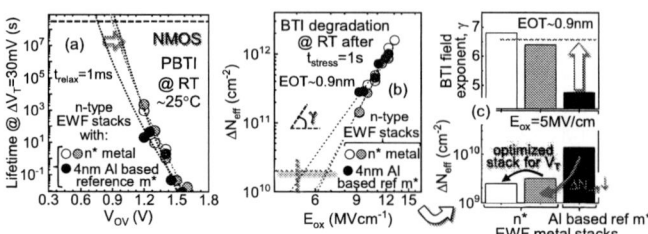

Figure 8. (a) Improved PBTI lifetime is extracted for GAA NS FETs with two vertically stacked nanosheets and built using an alternative, thinner n* EWF metal as opposed to the 4nm Al based m* reference gate stack process described in Figs. 4b,e. The first type of devices also exhibits lower ΔN_{eff} and higher γ values, indicative of an overall better reliability behavior by corresponding, in general, to a narrower distribution of charges/ defects in the gate dielectric *vs.* Al based reference transistors.

Reliability wise, and in good agreement with the noise results, Fig. 8 shows improved bias-temperature-instability (BTI) lifetime is extrapolated for devices with the n* EWF metal, for which lower effective charging defects density (ΔN_{eff}) and higher PBTI field exponent γ (also called voltage acceleration factor γ) values are extracted in Figs. 8b,c. Both are indicative of superior device reliability behavior, typically corresponding to a narrower distribution of charges/defects in the gate dielectric than for the reference.

Vertical Nanowire and Nanosheet FETs

Hybrid scaling is increasingly gaining traction wherein devices can be customized to meet specific system requirements, instead of relying on a single one-meets-all generic technology. In this context, the feasibility of co-integrating on the same wafer, and in a cost-effective way, finFETs (or vertically stacked lateral NS FETs, the most promising finFET replacements for sub-5nm nodes) for high-performance logic and GAA VNS or VNW FETs for specialized memory circuits, e.g., as an MRAM selector is evaluated by process simulations, experimentally validated for each type of devices. One such possible scheme is illustrated in Fig. 9, wherein cost adders were minimized by sharing the maximum number of critical process steps for both types of devices, such as: fin and VNS patterning; buried metal line formation; dummy gate patterning; epi growth (for S/D in finFETs; and on top of the pillars in VNS FETs for obtaining a larger contact area and lower S/D series resistance $R_{S/D}$); dummy gate removal, gate stack deposition and metal CMP at RMG module. Middle-of-line and BEOL are also common.

As schematically illustrated in Fig. 10, typically for finFET based MRAMs the array of magnetic tunnel junctions (MTJ) is embedded in the BEOL in-between two metal levels (38-42), such as M3 and M4 (39). GAA VNS FETs, on the other hand, allow an earlier introduction of the MTJ element in the flow, immediately after the top electrode (TE)'s formation, with the MTJ pillars being then placed on top of it. In this case, and in agreement with the process simulations in Fig. 9, the source and word lines (SL, WL) can be routed with a buried metal line defined after VNS patterning and via the gate electrode with M2, respectively, while the bit line (BL) is routed with M1. The BML is a key element in the cell design as it enables considerable routing simplification to access the bottom region of the vertical transistors. In Fig. 9 the latter consist of two vertical

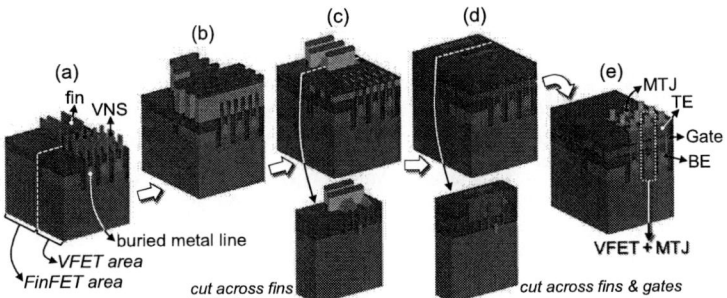

Figure 9. Schematics illustrative of the process simulations done of a cost-effective, co-integration flow to fabricate on the same wafer triple-gate finFETs and GAA VNS FETs. a) shows the common fin/VNS patterning step (two VNS per VFET device depicted) and buried power rail formation, with b) and c) representing the dummy gate patterning and epi growth (on the finFET's S/D and on top of the VNS) steps, respectively. Gate stack implementation is done here using an RMG scheme (d). In addition, when building the VNS FETs for use as MRAM selectors, the magnetic tunnel junction (MTJ) element can be introduced earlier on in the flow as compared to the case of finFET based cells, just after the top electrode (TE)'s formation (e).

nanosheets per device with a BML defined in-between them, its resistance set by the metal used for its definition and by its thickness.

The substantial gains in STT-MRAM's cell area reduction enabled by the use of vertical transistors are illustrated in Fig. 11, where cell layouts using finFET or VNS FET selectors and 3nm node's design rules are compared. Adopting square bitcells, the finFET selector consists of four active fins, corresponding to a cell size of 2 CGP × 2 CGP. A

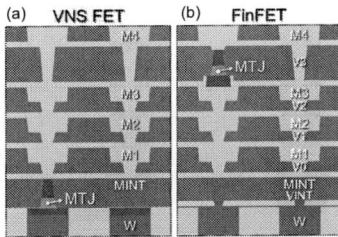

Figure 10. A schematic comparison of the location of the MTJ element for STT-MRAM cells using vertical NS FETs (a) vs. finFETs (b) as the cell selector transistors. Whereas for the latter, the MTJ pillars are inserted in the flow in-between metal 3 (M3) and metal 4 (M4) levels, for VNS FET based cells, they can be introduced earlier on in the flow, just after the top electrode (TE)'s formation.

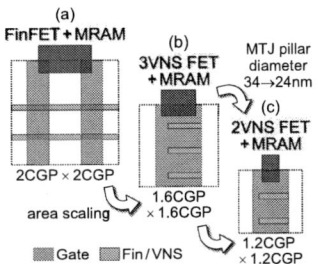

Figure 11. Assuming 3nm node's design rules, a comparison of STT-MRAM cells designed using as selectors either finFETs (consisting of 4 active fins per cell in (a)) or VNS FETs (with 3 or 2 VNS / cell in (b,c), respectively) shows considerable cell area reduction (a→b→c) is enabled with the use of vertical FETs. For 3→2 VNS / cell, the MTJ diameter was retargeted in design from 34 to 24nm through a DTCO loop.

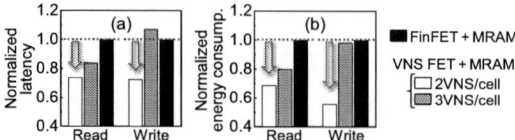

Figure 12. Considerably smaller access latency (a) and lower energy consumption (b) values are obtained for both read and write operations for a 2Mbit STT-MRAM macro designed with vertical NS FET instead of finFET selectors, especially for the more scaled VNS FET cell version.

36% area reduction is obtained when replacing finFETs by VNS FET selectors consisting of three VNS per device. Further scaling is achieved with two VNS per FET selector by retargeting the MTJ pillar diameter from 34nm to 24nm through a design technology co-optimization (DTCO) loop: 64% smaller cells are obtained than when using finFETs. Overall, as shown in Fig. 12, considerably lower read and write energy consumption and access latency values are extracted for a 2Mbit STT-MRAM macro designed with the more scaled 2VNS FET selector. This is due to the lower parasitic resistances and capacitances for the WL, SL and BL of these smaller cells, and thanks also to reduced VNS FET device parasitics (33,37,48).

Specific to vertical FETs' fabrication, two critical aspects need to be addressed for their reliable implementation: 1) the handling of an intrinsically non-self-aligned gate electrode towards S/D; 2) the feasibility of continuing using process-induced stress techniques for mobility boost.

Fig. 13 shows results from a TCAD evaluation of the detrimental impact of having the gate electrode vertically misaligned towards the top (D) or bottom (S) part of the pillars. Such vertical shifts of the gate position result in underlap or overlap at the S or D with considerable I_{ON} loss and I_{OFF} increase. Less I_{ON} impact is predicted for junctionless (JL) *vs.* inversion-mode (IM) FETs and for higher channel doping concentration values (N_{NW}), but the latter comes at the expense of larger variability and DC performance degradation (due to mobility loss) (34,35). Fig. 14 shows the schematics of a simple

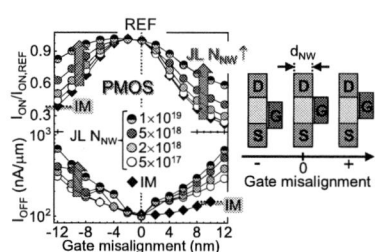

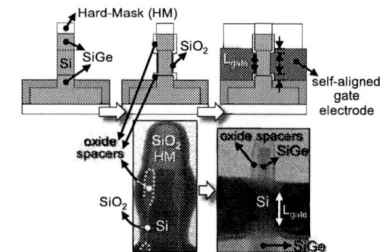

Figure 13. TCAD results for GAA VNW FETs show I_{ON} and I_{OFF} are impacted in case of vertical misalignment of the gate electrode towards the source and/or drain (data shown for V_{DS}=-1V, L_{gate}=30nm, d_{NW}=10nm, $N_{S/D}$=1×10^{20} at/cm^3).

Figure 14. Schematics and TEM images illustrating a simple flow to obtain self-aligned oxide spacers for GAA vertical NW or VNS FET devices. It relies on the different oxidation kinetics of Si and SiGe, using $Si_{0.75}Ge_{0.25}$/Si channel/$Si_{0.75}Ge_{0.25}$ VNW on VNS instead of Si-only pillars.

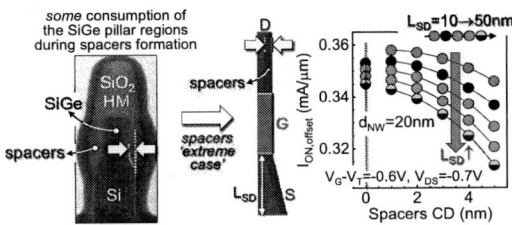

Figure 15. Some lateral consumption of the SiGe pillar regions is expected to occur during the self-aligned oxide spacers' formation process described in Fig. 14. TCAD simulations show, however, that the impact on I_{ON} can be mitigated with optimization of the SiGe regions' height and the spacers' thickness as shown in the figure on the right-hand side, without affecting C_{GG} (37) (data plotted for JL devices with $N_{NW}=5\times10^{17}at/cm^{3}$).

solution for a flow to obtain self-aligned oxide spacers by: 1) instead of Si-only pillars, use VNW or VNS consisting of a Si section (the channel) vertically neighbored by two highly doped SiGe (~25% Ge) regions; 2) rely on the different oxidation kinetics of Si and SiGe to create the oxide spacers surrounding the SiGe regions and to set their thickness, so that the gate electrode overlaps said spacers for a well-controlled L_{gate} as seen in the TEM images shown in Figs. 14 and 15 (35,37). During the spacers' formation process some lateral consumption of the SiGe in the pillars occurs, being less for thinner spacers. The simulation results in Fig. 15 show, however, that its impact on $R_{S/D}$ and hence DC can be mitigated by reducing the height of the SiGe regions and the targeted spacers' width or thickness (also called the spacers' critical dimension (CD)), with Ref. (37) confirming that no C_{GG} degradation by such structural tuning is obtained.

A replacement metal gate process, standardly used in finFETs and lateral NS FETs can also be advantageous for VNW or VNS FETs as it can allow shrinkage of the channel's cross-section (33) for improved electrostatics (without impacting the bottom/top S/D regions), while also enabling new techniques for stress-induced mobility enhancement (37). Feasibility of the latter is another important aspect to consider when evaluating the potential for implementation into manufacturing of this novel type of devices.

Fig. 16 shows a scheme wherein a SiGe stressor layer is epitaxially grown around the vertical Si channel after spacers formation, e.g., the self-aligned oxide spacers described in Figs. 14,15 (37). The stressor layer is later removed at RMG module, after dummy gate

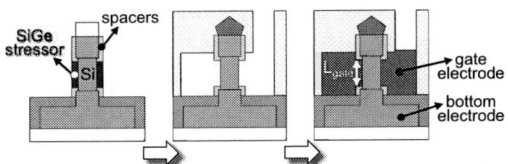

Figure 16. Illustration of a scheme to introduce stress in the vertical channel of n-type GAA VNW or VNS FETs for mobility enhancement (37): 1) a SiGe stressor is epitaxially grown around the Si channel after self-aligned spacers formation; 2) this layer is removed at the RMG module prior to the gate stack deposition, while the top of the pillars is kept encapsulated by a cap layer which remains connected to the bottom isolation layer at some places.

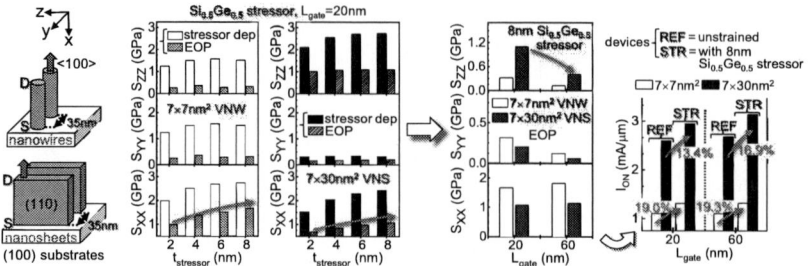

Figure 17. On the left-hand side, impact of the stressor's thickness on the average stress (X, Y, and Z components: S_{XX}, S_{YY}, S_{ZZ}) induced in the channel for n-type $7\times7nm^2$ VNW and $7\times30nm^2$ VNS FETs after stressor's deposition and at end-of-process (EOP). Overall, a thicker stressor leads to higher stress and larger I_{ON} values. The effect of the channel length on the average stress remaining in the channel at EOP is shown on the right-hand side plots. A stronger relaxation of the undesirable S_{ZZ} for longer L_{gate} in VNS FETs results in higher I_{ON} gains for taller devices (current values normalized by VNW/VNS pitch at $I_{OFF}=100nA/\mu m$, $V_G=V_{G,OFF}+0.7V$ for strained (STR) vs. unstrained devices (REF)).

removal and prior to the gate stack deposition. Stress memorization into its surroundings is enabled by the fact that the layer encapsulating the top part of the pillars remains connected to the bottom isolation layer at some places. This scheme was evaluated by stress and ballistic current simulations which were done considering the effects of surface quantum confinement, channel orientation and stress (37,43). Fig. 17 shows that, for both VNW and VNS n-type FETs, a tensile stress is introduced in the channel with the stressor's growth. Some amount of it remains in the channel after the stressor is removed (a stage called here 'at end-of-processing (EOP)'), depending on the structure dimensions. Overall, for both types of vertical devices, a thicker stressor ($2 \rightarrow 8nm$) induces higher stress values. However, while for a VNW both the Z and Y stress components relax at EOP, for a $7\times30nm^2$ VNS ~1GPa of tensile stress remains in the Z direction (S_{ZZ}). The latter is detrimental as it leads to degradation of the channel mobility. As a result, higher I_{ON} gains are predicted for VNW FETs. Interestingly, Fig. 17 also shows that a higher relaxation of the unwanted S_{ZZ} occurs for devices with longer L_{gate}. This allows a larger I_{ON} increase for taller VFETs, especially in the case of VNS FETs, as demonstrated by the ballistic currents reported in Fig. 17. They were calculated for VNW and VNS FETs with two different L_{gate}, noting that their values should be taken as indicative only, especially for the longest channels. Using a similar stressor, the I_{ON} improvement when increasing L_{gate} from 20nm to 60nm is here: 19.0% $\rightarrow$ 19.3% and 13.4% $\rightarrow$ 16.9% for VNW and VNS FETs, respectively. These results highlight that there is a smaller impact of the device's geometry on the I_{ON} gains for longer L_{gate} values.

Conclusions

Vertically stacked lateral NS FETs with a GAA configuration are shown to be the leading candidates to succeed finFETs and maintain the PPAC logic roadmap for advanced sub-5nm technology nodes. Implementation of a scaled vertical distance between the nanosheets in these devices and of thinner gate stacks were identified as key fabrication challenges for obtaining improved DC and AC device characteristics and were thus studied in-depth.

At the same time, with hybrid scaling increasingly gaining traction to achieve higher system value and flexibility, GAA vertical FETs are also interesting to explore thanks to their potential for enabling increased and more energy efficient on-chip memory content, e.g., as the selector transistors for STT-MRAM. For these novel devices, a scheme for self-aligned spacers has been evaluated to avoid that vertical misalignments of the gate electrode towards the source/drain impact the device's performance, while also improving its variability control. The feasibility to introduce stress for mobility enhancement in the vertical channels has also been demonstrated, exploring dependencies on materials and device geometries.

Acknowledgments

The imec CMOS core partners, the European Commission and local authorities, and the imec 300mm pilot line are acknowledged for their support.

References

1. S. C. Song *et al.*, *IEEE VLSI Tech. Dig.*, p. 198 (2015);
2. L. Liebmann *et al.*, *IEEE VLSI Tech. Dig.*, p. 112 (2016);
3. M. Garcia Bardon *et al.*, *IEEE IEDM Tech. Dig.*, p. 687 (2016);
4. H. Mertens *et al.*, *IEEE IEDM Tech. Dig.*, p. 828 (2017);
5. M. Bohr, *IEEE IEDM Tech. Dig.*, p. 1 (2011);
6. K. J. Kuhn, *IEEE Trans. Elec. Dev.*, **59**(7), p. 1813 (2012);
7. M. De Marchi *et al.*, *IEEE IEDM Tech. Dig.*, p. 183 (2012);
8. S. Bangsaruntip *et al.*, *IEEE IEDM Tech. Dig.*, p. 526 (2013);
9. S.-G. Hur *et al.*, *IEEE IEDM Tech. Dig.*, p. 649 (2013);
10. A. Veloso *et al.*, *IEEE VLSI Tech. Dig.*, p. 138 (2015);
11. I. Lauer *et al.*, *IEEE VLSI Tech. Dig.*, p. 140 (2015);
12. H. Mertens *et al.*, *IEEE VLSI Tech. Dig.*, p. 142 (2015);
13. H. Wu *et al.*, *IEEE IEDM Tech. Dig.*, p. 16 (2015);
14. A. Veloso *et al.*, *IEEE VLSI Tech. Dig.*, p. 138 (2016);
15. H. Mertens *et al.*, *IEEE VLSI Tech. Dig.*, p. 158 (2016);
16. S. Barraud *et al.*, *IEEE IEDM Tech. Dig.*, p. 464 (2016);
17. H. Mertens *et al.*, *IEEE IEDM Tech. Dig.*, p. 524 (2016);
18. D. Jang *et al.*, *IEEE Trans. Elec. Dev.*, **64**(6), p. 2707 (2017);
19. L. Witters *et al.*, *IEEE VLSI Tech. Dig.*, p. 194 (2017);
20. N. Loubet *et al.*, *IEEE VLSI Tech. Dig.*, p. 230 (2017);
21. S. Barraud *et al.*, *IEEE IEDM Tech. Dig.*, p. 677 (2017);
22. Y. M. Lee *et al.*, *IEEE IEDM Tech. Dig.*, p. 681 (2017);
23. E. Capogreco *et al.*, *IEEE VLSI Tech. Dig.*, p. 193 (2018);
24. M. J. H. van Dal *et al.*, *IEEE IEDM Tech. Dig.*, p. 492 (2018);
25. R. Ritzenthaler *et al.*, *IEEE IEDM Tech. Dig.*, p. 508 (2018);
26. C. W. Yeung *et al.*, *IEEE IEDM Tech. Dig.*, p. 652 (2018);
27. E. Capogreco *et al.*, *IEEE VLSI Tech. Dig.*, p. 94 (2019);
28. A. Veloso *et al.*, *SSDM Tech. Dig.*, p. 559 (2019);

29. R. Bao *et al.*, *IEEE IEDM Tech. Dig.*, p. 234 (2019);
30. N. Loubet *et al.*, *IEEE IEDM Tech. Dig.*, p. 242 (2019);
31. J. Zhang *et al.*, *IEEE IEDM Tech. Dig.*, p. 250 (2019);
32. T. Huynh-Bao *et al.*, *SPIE Proc.*, **9781**, 978102 (2016);
33. A. Veloso *et al.*, *ECS Trans.*, **72**(4), p. 31 (2016);
34. A. Veloso *et al.*, *SSDM Tech. Dig.*, p. 221 (2017);
35. A. Veloso *et al.*, *SSDM Tech. Dig.*, p. 159 (2018);
36. T. Huynh-Bao *et al.*, *ACM/IEEE DAC Proc.*, 267-TY328, s13p1 (2019);
37. A. Veloso *et al.*, *IEEE IEDM Tech. Dig.*, p. 230 (2019);
38. O. Golonzka *et al.*, *IEEE IEDM Tech. Dig.*, p. 412 (2018);
39. S. Sakhare *et al.*, *IEEE IEDM Tech. Dig.*, p. 420 (2018);
40. W. J. Gallagher *et al.*, *IEEE VLSI Tech. Dig.*, p. 190 (2019);
41. S. Aggarwal *et al.*, *IEEE IEDM Tech. Dig.*, p. 18 (2019);
42. K. Lee *et al.*, *IEEE IEDM Tech. Dig.*, p. 22 (2019);
43. Synopsys Sentaurus Device, Process & Monte Carlo Band Structure user guides (2018);
44. S. M. Y. Sherazi *et al.*, *SPIE Proc.*, **10148**, 101480Y (2017);
45. A. Gupta *et al.*, *IEEE IITC Tech. Dig.*, p. 4 (2018);
46. E. Simoen *et al.*, *ECS J. Solid State Sci. and Tech.*, **3**(6), p. Q127 (2014);
47. W. Fang *et al.*, *IEEE Elec. Dev. Lett.*, **37**(4), p. 363 (2016);
48. A. V.-Y. Thean *et al.*, *IEEE VLSI Tech. Dig.*, p. 26 (2015).

Fully-Depleted SOI Technology For High-Power RF Applications

T. V. Dinh[a], B. W. C. Hovens[b], M. Vroubel[b], I. Brunets[b], H. P. Tuinhout[c], L. F. Tiemeijer[c], N. Wils[c], G. T. Sasse[b], P. Grudowski[d], M. Raucoules-aime[e], S. Dal Toso[e], M. Cannella[e] and C. Ghidini[e]

[a] NXP Semiconductors, Leuven, Belgium
[b] NXP Semiconductors, Nijmegen, The Netherlands
[c] NXP Semiconductors, Eindhoven, The Netherlands
[d] NXP Semiconductors, Austin, TX, USA
[e] NXP Semiconductors, Mougins-Sophia Antipolis, France
correspondent author email: thanhviet.dinh@nxp.com

> This paper analyzes the latest development of critical RF devices which are capable of delivering both high RF performance and high-voltage robustness in fully-depleted SOI technology. The unique features of RF measurements and EM simulations for these devices, together with their impact on associated designs of RF applications, are discussed.

Introduction

RF connectivity is the key enabler for Internet of Things and for transferring a continuously-increasing amount of multimedia data. Cost-effective high-performance RF front-end ICs are needed to facilitate such RF connections, which range from low frequency / low power (BLE / Zigbee, 2.4GHz) via high frequency / high power (WiFi, 2.4-5GHz, 802.ac / 802.ax), up to higher frequency (5G, 28GHz) for much larger bandwidth. In a much higher frequency range (77GHz and 140GHz), high-resolution millimeter wave (mmW) radars are integrated with powerful microprocessors to realize a higher level of autonomous driving (1). The demand for a higher bandwidth, higher data rate, more power efficiency and better resolution for these various RF systems poses challenges for both system design and RF capability of the adopted technologies.

Although III-V and SiGe technologies are traditionally, and still, dominant in high-end RF systems (2-4), the requirement of smaller footprint, higher level integration of signal processing, together with the enhanced RF performance in advanced CMOS nodes over the last years, has opened opportunities for CMOS-based RF systems. The applications designed on RF-CMOS have been indeed extended from low-end low-power to high-end complex systems, e.g. RF-CMOS based car radars (1). This paper will review and analyze the key figures of merit of RF devices which are critical to RF transceivers in an advanced Fully-depleted Silicon on Insulator (FDSOI) process, both in the regime of small-signal and large-signal applications.

Compared to other planar bulk technologies at the same node, FDSOI has shown its benefits in low power applications by a better electrostatic control, a reduction in leakage currents and device variability, which enables a lower minimum required supply voltage (V_{DD}) (5-6). Figure 1 shows the cross-section of an MOS device in an advanced FDSOI process. Thanks to the isolation box in the substrate and its intrinsic channel, the parasitic capacitances are also reduced significantly, including the capacitance between source and drain, and between drain and body. These features help to boost the transistors' RF performance. The published data shows the measured f_T (cut-off frequency) and f_{MAX} (maximum oscillation frequency) of RFMOS in FDSOI processes reach 350 GHz (7). Moreover, by the process construction (gate-first vs. gate-last), the gate resistance of the transistors in FDSOI processes is smaller (7), which is beneficial to their RF noise performance. These features are important for small-signal RF blocks in Front-end modules, e.g. LNA, VCO and mixers. The first part of this paper reviews the key figures of merit for RFMOS fabricated in FDSOI and other comparable technologies.

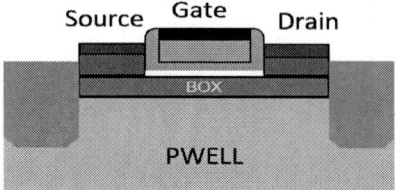

Figure 1. Schematic of the cross-section of an NMOS in sub-28nm FDSOI processes.

In addition to small-signal applications, FDSOI capability for the transmitter part (TX) including integrated power amplifiers (PA) for generating a higher power level is also explored. In previous mobile generations (3G/4G), the base station needs to deliver peak power more than 100W, which are solely enabled by GaN or GaAs technologies, and their corresponding handsets also need to transmit a few Watts, which are out of the reach of CMOS-based PA. By moving to massive MIMO with a large number of antennas, the power per RF PA is reduced significantly for both base-stations and handsets in 5G network, with the P_{MAX} per PA varying from 20 dBm to 25 dBm (2-3). Such sub-Watt power range opens the opportunities for making silicon-based RF PA, enabling a much larger scale of RF Front-end modules integration. To make the FDSOI process possible for high-power RF applications including 5G and beyond, the development of critical active and passive components of high-breakdown and high RF-performance is essential. The second part presents the device architecture, and electrical DC and RF performance of RF-LDMOS devices which can deliver the expected performance. The passive devices, including fringe capacitors and transformers, are discussed in details in the last section of this paper.

To facilitate the development of such passive devices, the validation of EM (electromagnetic) simulation is critical. Therefore, in the last section, the agreement between EM simulation and RF measurement for passive devices is also discussed. In addition, the 4-port test structure designs and corresponding in-house 4-port RF characterization (22) to measure 4-terminal devices such as the transformers or the RFMOS with back-gate bias are presented. The paper is concluded with the design perspectives for RF applications based on the standard and newly developed devices in an FDSOI process.

RF-CMOS performance in FDSOI process

The RF performance of NMOS devices has been improved significantly over the last years. Although their performance is still lagging behind compared with state-of-the-art SiGe HBTs, they are sufficient for a wide range of RF applications, up to 77GHz for radar sensors (1). One of the key figures of merit to measure such performance is cutoff frequency (f_T) and maximum oscillation frequency (f_{MAX}), which can be expressed by the following first-order equations (8):

$$f_T = \frac{1}{2\pi} \cdot \frac{g_m}{C_{gg} + C_{par} + C_{ov}},$$ [1]

$$f_{MAX} = \frac{f_T}{2\sqrt{(R_g + R_i)(g_{ds} + \pi \, f_T C_{ov})}},$$ [2]

where g_m is the trans-conductance, C_{gg}, C_{par}, C_{ov} are the input capacitance, parasitic gate-bulk capacitance, gate-drain overlap capacitance respectively, R_g is the gate resistance, R_i is the real part of the input impedance and g_{ds} is the output conductance. Hence, a large g_m together with low input and overlap capacitances are needed to have a high f_T. For achieving a high f_{MAX}, in addition to a high f_T, the gate resistance should be as small as possible. Figure 2 shows the f_T and f_{MAX} of the RF-NMOS from different CMOS technologies, including bulk, FDSOI and FinFET.

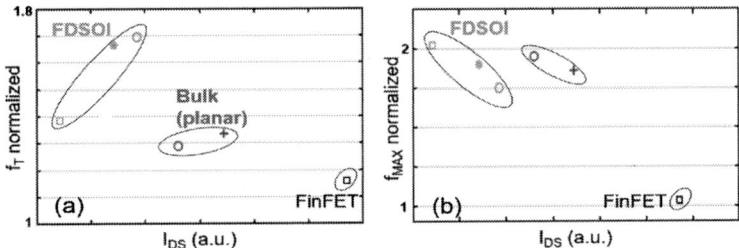

Figure 2. The benchmarking of maximum f_T (left) and f_{MAX} (right) of RF-NMOS from different CMOS technologies (5-7, 9-13). The parasitics from interconnects up to Metal 3 are taken into account.

Figure 2 shows that both f_T and f_{MAX} of RF-NMOS in FDSOI processes are higher than the corresponding figures from the devices on bulk in a similar CMOS node. Moreover, FDSOI NMOS needs a smaller drain current to reach the peak f_T and f_{MAX}, which is very beneficial for low-power RF applications. Although FinFET has the best electrostatic control and a higher transconductance, it has higher parasitic capacitances (from the gate to the drain / source) and therefore lower f_T and f_{MAX} (9-13).

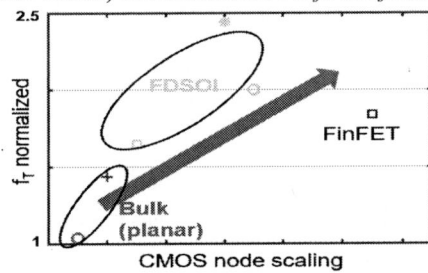

Figure 3. The trend of f_T over CMOS nodes scaling for different technologies (planar, FDSOI, FinFET).

In addition to the maximum, the value of f_T and f_{MAX} at $g_m/I_{DS}=10$ is another important merit indicating the small-signal gain capability of the technology, which is important for RF and analog designs . Figure 3 shows the trend of f_T over the CMOS nodes scaling. A smaller node helps higher g_m, and therefore a higher f_T.

RF high-voltage active devices in FDSOI

The availability of high-performing, high-voltage capable RF devices is required to enable integrated power amplifier (PA) designs. Figure 4 shows the schematic of a common RF PA for WiFi applications. For the highest efficiency and lowest cost, in this design, the PA is directly attached to the battery (3.6V or 4.8V) with integrated TX driver, meaning a 5V device is needed. Lateral-diffused MOS (i.e. LDMOS) or Extended-drain MOS (i.e. EDMOS) are the most suitable for this purpose.

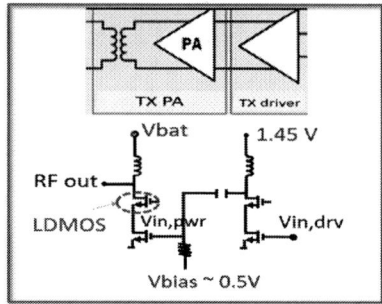

Figure 4. Schematic of an integrated battery-attached PA with TX driver (for the highest efficiency and lowest cost) for WiFi (14).

Although 5V-LDMOS transistors are developed and offered in most advanced processes – ranging from bulk to FDSOI (15-19) and from planar to FinFET (9, 20-21) - their RF performance is often quite limited. As well-known in RF circuit design, for most implementations, the cutoff frequency f_T of the device must be at least 5x higher than the operating frequency (2-3). This means that the RF-LDMOS should have $f_T \sim$ 30GHz for WiFi, and $\sim$ 100GHz for 5G mobile. Reaching this range of f_T's while meeting the reliability requirement for the usage of 3.3V or 5V is a big challenge. Moreover, to ensure cost-effectiveness and portability from one process to another, such a device should preferably be made without using additional dedicated masks or implants. This proves impossible for the current LDMOS architectures. In (14), we have shown a novel 3.3V / 5V RF-LDMOS with f_T above 100GHz, which have been fabricated in an advanced sub-28nm node FDSOI process flow. The devices are implemented in the non-SOI area.

Device Architecture

The design of the 5V LDMOS follows the concept of an EDMOS with a record high f_{MAX} in a baseline 40nm CMOS process presented in (19) (see Figure 5). In standard LDMOS devices, a thick and uniform oxide in the channel and in the drain extension are used to handle a large drain-gate voltage. Therefore, the scaling of the channel length is limited by gate-oxide thickness (IO gate oxide is typically used), which means that the trans-conductance g_m cannot be increased significantly.

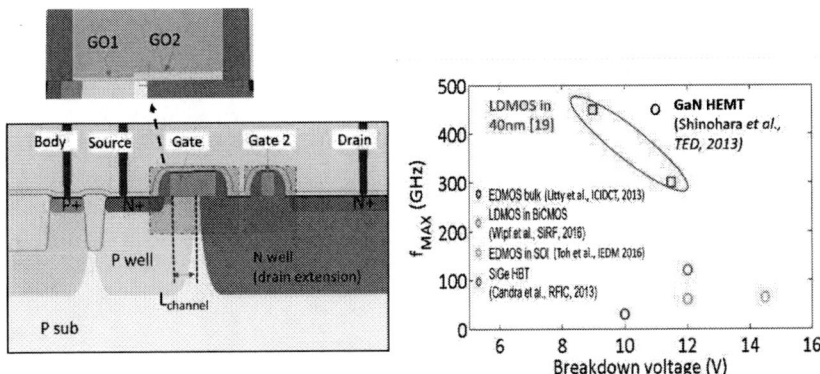

Figure 5. Cross-section of the LDMOS developed in 40nm, which has a record performance in both f_T and f_{MAX} (19).

The device proposed in (19) has the following unique features to boost f_T and f_{MAX}: a) a thin gate-oxide (logic) in the channel to allow the channel length scaling down to the minimum length of the process b) a thick gate-oxide (IO) in the drain extension lightly doped by an available N-well implant to accommodate the high electric field occurring at the gate-drain edge c) the gate-drain overlap is reduced significantly to reduce C_{ov}. To circumvent the hot carrier injection resulting from such a reduction, a dummy gate (i.e. Gate 2 in Figure 5) is formed to reduce the electric field at the gate-drain edge. Moreover, such a split gate has advantages over a long gate-drain overlap in term of the capacitance reduction and the flexibility in the biasing scheme. Indeed, it can be connected to the main gate, biased independently or floated depending on its effectiveness in improving the device reliability.

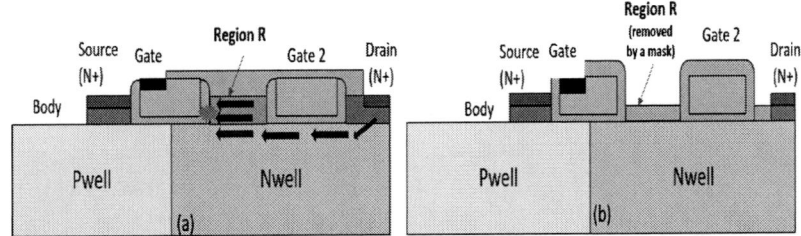

Figure 6. (a) The LD(N)MOS ported (as-is) from 40nm bulk into FDSOI process with raised epi-S/D process. (b) Conventional way of making the LDMOS in raised S/D process: region R is removed by using a mask (14).

However, the architecture of such an LDMOS has implementation issues due to the raised epitaxial growth (Source / Drain formulation) in sub-28nm FDSOI processes. For LDMOS operating at higher voltages, such a raised S/D creates a very high electric field between the gate-edge and the drain extension, causing high gate-drain leakage currents while increasing the risks for gate oxide breakdown, see Figure 6a. To circumvent this, additional layers and masking steps are conventionally used to block the epi growth in the drift region. This blocking removes the direct path of the current and hence the flow of hot carriers (i.e. carriers with high energy) in the high-field region. However, this solution (Figure 6b) adds costs to the baseline technology offerings.

The high-performance RF-LDMOS has been constructed by a novel usage of multiple dummy gates in the drift region, in combination with an LDD implant of the opposite dope type (i.e. PLDD for LDNMOS and NLDD for LDPMOS). The device cross-sections are shown in Figure 7. With this approach, the epi-growth is partly blocked, and an island with an opposite dope type is formed between the main gate and the dummy gate. Such a drain extension construction will push the drain current path away from the surface and from the gate-drain edge, reducing device leakage while improving reliability performance.

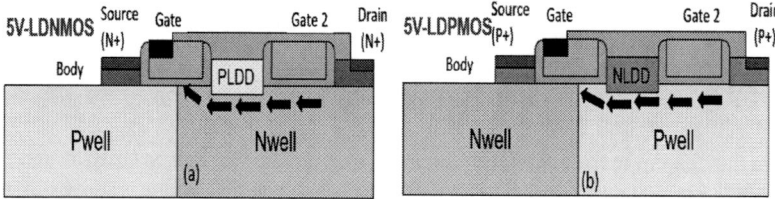

Figure 7. Cross-sections of the 5V-LDMOS developed in FDSOI process for RF high-power applications (14).

The 3.3V LDMOS is designed to reach an f_T above 100GHz, which makes it suitable for 5G mobile applications. To achieve that, a more aggressively scaled channel length is required. Therefore, the GO1 gate oxide is used in the channel and the GO2 gate oxide in the drain extension, which is shown in Figure 8. The rest is the same as in the 5V device.

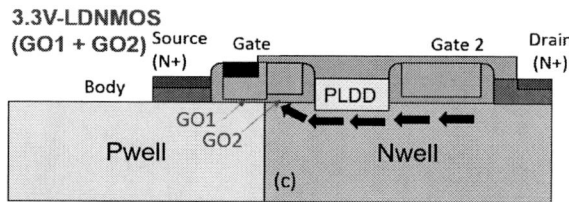

Figure 8. Cross-section of the 3.3V LDMOS aimed for f_T of 100GHz (14).

<u>DC and RF Characteristics</u>

DC and RF performance of the 5V-LDMOS are shown in Figure 9, surpassing key aspects of the device implemented in 40nm technology. In term of DC characteristics, the device in FDSOI has similar linear I_{DS} and much better leakage current. This device also delivers higher peak f_T. Moreover, due to a smaller node in this FDSOI process, the gate length in GO2 device can be scaled down much further without the need of using a combination of core gate oxide GO1 (for the channel) and IO oxide GO2 (for the drain extension). Therefore, the device in this FDSOI process has a higher maximum gate voltage due to a uniform GO2 gate oxide, providing more headroom in gate and drain biasing for RF PA design.

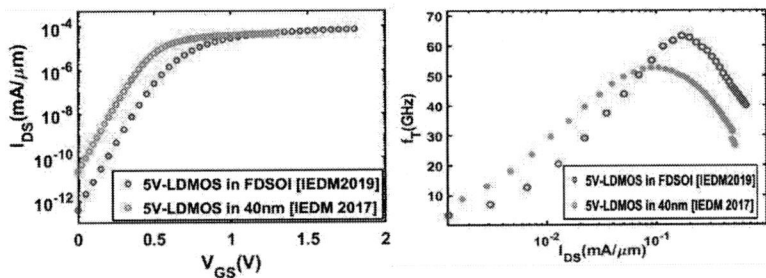

Figure 9. DC and RF performance of 5V-LDMOS in FDSOI (*IEDM2019*, (14)), in comparison with the similar device developed in bulk 40nm process (*IEDM2017*, (19)).

The f_T of the 3.3V-LDMOS with a channel length of sub-100nm is shown in Figure 10a. The achieved $f_T > 100$GHz (de-embedded to Metal 2) is significantly higher than f_T of a 1.8V GO2 RF-NMOS in the same process, meaning this 3.3V RF-LDMOS is much more suitable for PA designs in 28GHz. The benchmarking of f_T vs. breakdown voltage in Figure 10b shows that f_T of the devices developed in this FDSOI process could be higher than high-voltage SiGe HBTs optimized for PA applications (24), and even comparable to state-of-the-art GaAs pHEMT used for 5G designs (f_T of 120GHz and nominal voltage of 4V) (4, 25).

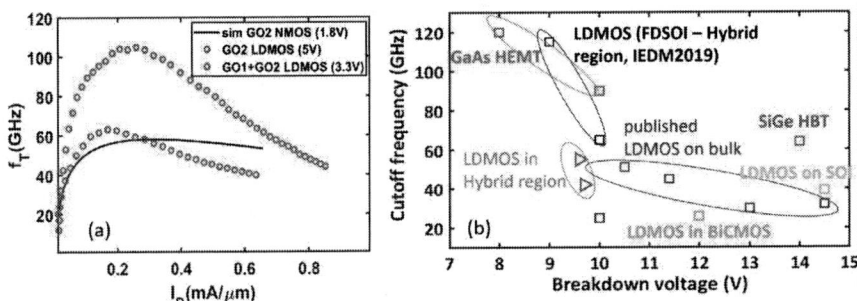

Figure 10. (a) Measured f_T of 3.3V RF-LDMOS device; (b) Benchmark of the devices targeted for 3.3V and 5V usage in various processes.

RF high-voltage passive devices in FDSOI

Fringe Capacitors

Device architecture. Metal-Oxide-Metal (MOM) capacitors are used in a number of different use cases and frequency ranges in connectivity applications. This type of capacitor is well suitable for high precision purposes because of their high capacitance density, small temperature and voltage dependence and good mismatch characteristics. The importance of specific figures of merit of these devices depends on the use case. Analog-digital converters operating at ~100MHz require high-precision capacitances (< 1fF), and excellent matching performance. For VCO's operating at GHz range, apart from a good precision and matching, the quality factor of the capacitor is a key parameter. In this paper, the architecture and RF characteristics of MOM developed for power amplifiers are presented. The devices are capable for 7V operation with a high quality factor.

The BEOL configuration used in this development comprises 6 thin copper, 2 thick copper and one aluminum metal layers. The first two thin metal layers (M1 and M2) are double patterned. Next thin metal layers (M3-M6) have alternating preferred directions, allowing finer metal pitch for preferred direction than for non-preferred direction. This will impose constraints on the construction of the MOM capacitors.

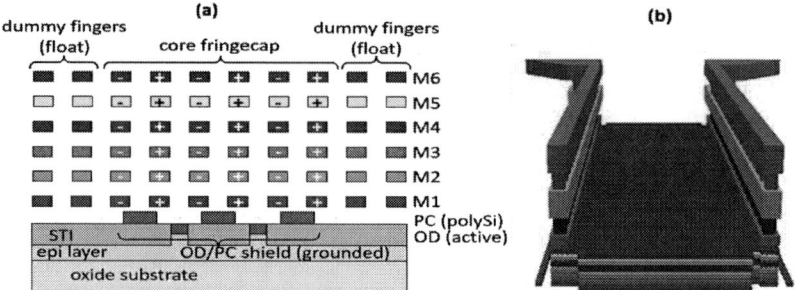

Figure 11. (a) Cross-section and (b) 3D view of the high-voltage fringe capacitors.

The architecture of the MOM capacitor is shown in Figure 11. Parallel fingers are in subsequent metal layers with equal polarity in vertical direction and alternating polarity in horizontal direction. Two side taps connect the fingers of positive and negative polarity, respectively. Optionally, electrically non-connected dummy fingers are placed at the ends of the MOM capacitor for improving the structural homogeneity for manufacturing process. Finger length, number and spacing and amount of metal layers involved are variable in order to select the desired capacitance and voltage capability. Usually, a grounded shield of alternate active (OD) and poly (PC) stripes is included underneath the capacitor. The minimum finger spacing is determined by the minimum metal spacing allowed in the manufacturing process and has a directional dependency. In preferred direction, the minimum spacing is 0.04μm while in non-preferred direction this spacing is 0.08μm. Given the architecture of the MOM capacitor and the preferred direction alternating for subsequent metals, the minimum finger spacing of 0.08μm is chosen.

Device performance. The RF performance of these capacitors have been measured up to 50GHz, and the open-short-load de-embedding procedure is followed to extract the final results, in the format of S or Y parameters (22). Then, the capacitance C_S, series resistance R_S and quality factors Q_S are calculated by:

$$C_s = \frac{1}{2\pi f * Im\left(\frac{2}{Y_{12} + Y_{21}}\right)} \tag{3}$$

$$R_s = Re\left(\frac{2}{Y_{12} + Y_{21}}\right) \tag{4}$$

$$Q_s = Re\left(\frac{1}{2\pi f R_s C_s}\right) \tag{5}$$

The architecture with parallel fingers in subsequent metal layers as shown in Figure 11 indeed protects the device from breakdown in vertical direction. Robustness at 7V operation is guaranteed by choosing a sufficiently large finger spacing in horizontal direction. Although a large finger spacing leads to a lower capacitance density, it helps to reach a higher Q-factor and better matching performance. Such a relation between Q-factor and capacitance density can be expressed by

$$Q = \frac{1}{\omega RC}$$

$$\approx \frac{1}{\omega * (\tan(\delta_D) + R_{sh} C_{den} \left(\frac{4fl^2}{3fw} + \frac{2(fw + fs) * fl * nf^2}{3tap_w}\right))} \qquad [6]$$

where ω is the angular frequency, R_{sh} is the metal sheet resistance, C_{den} is the capacitance density and tap_w is side tap width. From equation 6, for the same capacitance value, the fringe capacitor with a lower capacitance density has a lower resistance, and hence, higher quality factor, when compared to a similar device with a higher capacitance density.

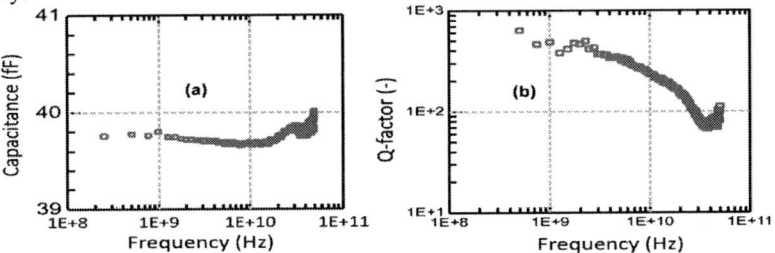

Figure 12. Measured capacitance and quality factor of the developed 7V fringe capacitor.

Figure 12 shows the capacitance (after being de-embedded) of the 7V fringe capacitor which has been developed to be used in WiFi PA. The capacitance is constant (~ 40fF) over frequencies as expected, and the device reaches a Q-factor larger than 100 at 28GHz, and 340 at 5GHz, which is state-of-the-art for fringe capacitors (23).

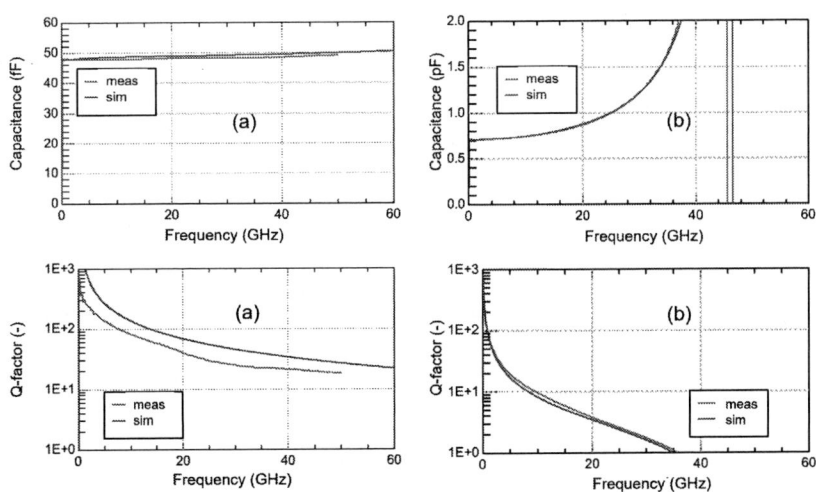

Figure 13. EM simulations vs RF measurements for two fringe capacitors of very different capacitance: (a) fF range and (b) pF range.

Electromagnetic (EM) simulation. The accuracy of EM simulation in predicting RF characteristics of passive devices is crucial for a right RF design. Within the scope of this work, an intensive calibration and validation for EM simulations have been done, by comparing the simulation results with RF measurements over a wide range of frequencies for different device types, including capacitors, inductors, transformers, and for different device geometries.

Figure 13 shows the comparison of the capacitance and quality factor between EM simulations and RF measurements for two fringe capacitors of very different capacitance, one at the range of fF and the other at pF. A very good matching between simulations and measurements is obtained for both devices, and for both capacitance and quality factor (i.e. series resistance). This verification assures EM simulations could be used reliably to optimize the designs for all passive components. The RF characteristics and EM simulations for the developed transformers are shown in the next section.

Transformers

Within the scope of this work, a two-way transformer has been developed to enable the design of an integrated PA for WiFi applications (circuit schematic in Figure 4). The device was realized in 3 top metal layers (2 copper and 1 aluminum), with the top view shown in Figure 14a.

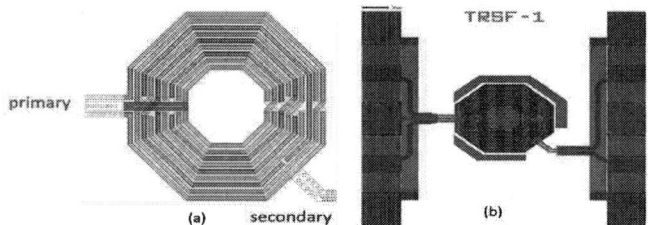

Figure 14. Layout of the transformer designed in 3 top metal layers: (a) top view (b) device placed in GSGSG pads for RF characterization.

The transformer has been placed in a 4-port GSGSG for RF measurement (22), as shown in Figure 14b. Figure 15 shows the measured performance of the transformer, with an excellent matching with EM simulation. A Q-factor of 12, at 5GHz, is sufficient for PA design, and further device optimization is possible.

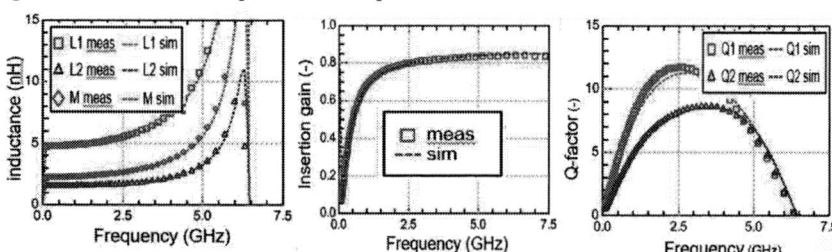

Figure 15. RF measurement vs. EM simulation for key figures of merit (inductance, insertion loss and Q-factor) of the developed transformer.

Conclusion

The RF performance of active and passive devices in an advanced FDSOI process has been analyzed and benchmarked. The combination of such high-performing standard devices thanks to the intrinsic construction of FDSOI process and the dedicated development of state-of-the-art high-voltage robust RF devices enables the highest level of integration for RF Front-end modules of various connectivity applications, ranging from WiFi, 5G to car radars and beyond.

References

1. L. Reger, *International Solid-State Circuits Conference*, p. 29 (2016)
2. D. Y. C. Lie, J. C. Mayeda, Y. Li, and J. Lopez, *Wireless Communications & Mobile Computing*, **2008**, 1, Wiley (2018).
3. P. M. Asbeck, N. Rostomyan, M. Özen, B. Rabet, and J. A. Jayamon, *IEEE Trans. Microwave Theory & Techniques,* **67** (7), 3099 (2019).
4. D. P. Nguyen, B. L. Pham, and A-V Pham, *IEEE Trans. Microwave Theory & Techniques*, **67**, 205 (2019).
5. N. Planes, O. Weber, V. Barral, S. Haendler, D. Noblet, D. Croain, M. Bocat, P.-O. Sassoulas, X. Federspiel, A. Cros, A. Bajolet, E. Richard, B. Dumont, P. Perreau, D. Petit, D. Golanski, C. Fenouillet-Béranger, N. Guillot, M. Rafik, V. Huard, S. Puget, X. Montagner, M.-A. Jaud, O. Rozeau, O. Saxod, F. Wacquant, F. Monsieur, D. Barge, L. Pinzelli, and M. Mellier, *Symposium on VLSI Technology*, p 133 (2012).
6. R. Carter, J. Mazurier, L. Pirro, J-U. Sachse, P. Baars, J. Faul, C. Grass, G. Grasshoff, P. Javorka, T. Kammler, A. Preusse, S. Nielsen, T. Heller, J. Schmidt, H. Niebojewski, P-Y. Chou, E. Smith, E. Erben, C. Metze, C. Bao, Y. Andee, I. Aydin, S. Morvan, J. Bernard, E. Bourjot, T. Feudel, D. Harame, R. Nelluri, H.-J. Thees, and L. M-Meskamp, *IEEE International Electron Devices Meeting (IEDM)*, p 27 (2016).
7. S. N. Ong, S. Lehmann, W.H. Chow, C. Zhang, C. Schippel, L.H.K. Chan, Y. Andee, M. Hauschildt, K.K.S. Tan, J. Watts, C.K. Lim, A. Divay, J.S. Wong, Z. Zhao, M. Govindarajan, C. Schwan, A. Huschka, A. Bellaouar, W. LOo, J. Mazurier, C. Grass, R. Taylor, K.W.J. Chew, S. Embabi, G. Workman, A. Pakfar, S. Morvan, K. Sundaram, M. T. Lau, B. Rice, *IEEE Radio Frequency Integrated Circuits Symposium (RFIC)*, p. 72 (2018).
8. P. H. Woerlee, M. J. Knitel, R. v. Langevelde, D. B. M. Klaassen, L. F. Tiemeijer, A. J. Scholten, and A. T. A. Z-V. Duijnhoven, *IEEE Trans. Electron Devices*, **48** (8), p. 1776 (2001).
9. J. Singh, C. Jerome, A. Wei, R. Miller, B. Arnaud, C. Lili, H. Zang, P. Kasun, P. Manjunatha, S. Biswanath, A. Kumar, S. M. Pandey, N. M. Iyer, A. Mittal, R. Carter, L. Zhao, E. Manfred, and S. Samavedam, *Symposium on VLSI Technology*, p. T140 (2014).
10. B. Sell, B. Bigwood, S. Cha, Z. Chen, P. Dhage, P. Fan, M. Giraud-Carrier, A. Kar, E. Karl, C.-J. Ku, R. Kumar, T. Lajoie, H.-J. Lee, G. Liu, S. Liu, Y. Ma, S. Mudanai, L. Nguyen, L. Paulson, K. Phoa, K. Pierce, A. Roy, R. Russell, J. Sandford, J. Stoeger, N. Stojanovic, A. Sultana, J. Waldemer, J. Wan, and W. Xu, *IEEE International Electron Devices Meeting (IEDM)*, p. 685 (2017).

11. H.-J. Lee, S. Morarka, S. Rami, Q. Yu, M. Weiss, G. Liu, M. Armstrong, C. - Y. Su, D. Ali, B. Sell, and Y. Zhang, *IEEE International Electron Devices Meeting (IEDM)*, p. 316 (2018).
12. K. W. J. Chew, A. Agshikar , M. Wiatr , J. S. Wong, W. H. Chow, Z. Liu, T. H. Lee, J. Shi, S. F. Lim, K. Sundaram, L. H. K. Chan, C. H. M. Cheng, N. Sassiat, Y. K. Yoo, A. Balijepalli, A. Kumta, C. D. Nguyen , R. Illgen , A. Mathew, C. Schippel, A. Romanescu, J. Watt, and D. Harame, *IEEE Radio Frequency Integrated Circuits Symposium (RFIC)*, p. 43 (2015).
13. B. Parvais, G. Hellings, M. Simicic, P. Weckx, J. Mitard, D. Jang, V. Desphpande, B. van Liempc, A. Veloso, A. Vandooren, N. Waldron, P. Wambacq, N. Collaert, D. Verkest, *European Solid-State Device Research Conference*, p. 158 (2018).
14. T. V. Dinh, B. W. C. Hovens, M. Vroubel, I. Brunets, H. P. Tuinhout, L. F. Tiemeijer, N. Wils, G. T. Sasse, P. Grudowski, M. Raucoules-aime, S. Dal Toso, M. Cannella, and C. Ghidini, *IEEE International Electron Devices Meeting (IEDM)*, p. 602 (2019).
15. M. Zierak, N. Feilchenfeld, C. Li, and T. Letavic, *IEEE International Symposium on Power Semiconductor Devices & IC's (ISPSD)*, p. 337 (2015).
16. R.-T. Toh, S. Parthasarathy, T. Sun, S. Zhang, R. V. Purakh, C. S. Zhu, V. S. Nune, J. S. Wong, M. Govindarajan, Y. K. Yoo, K. W. Chew, and D. S. Ang, *IEEE International Electron Devices Meeting (IEDM)*, p. 35 (2016).
17. C. Schippel, S. Lehmann, S.N. Ong, A. Muehlhoff, I. Cortés, W.H. Chow, D. Utess, A. Zaka, A. Divay, A. Pakfar, J. Faul, J. Mazurier, K.W. Chew, L.H.K. Chan, R. Taylor, B. Rice, and D. Harame, *IEEE SOI-3D-Subthreshold Microelectronics Technology Unified Conference (S3S)*, (2018).
18. A. Litty, S. Ortolland, D. Golanski, and S. Cristoloveanu, *IEEE International Symposium on Power Semiconductor Devices & IC's (ISPSD)*, p. 73 (2015).
19. T. V. Dinh, J. Sonsky, J. Claes, O. Dieball, G. T. Sasse and Celine Detcheverry, *IEEE International Electron Devices Meeting (IEDM)*, p. 621 (2017).
20. B. S. Kumar, M. Paul, M. Shrivastava, and H. Gossner, *IEEE International Symposium on Power Semiconductor Devices & IC's (ISPSD)*, p. 72 (2018).
21. Y.-T. Wu, F. Ding, D. Connelly, P. Zheng, M.-H. Chiang, J. F. Chen, and T.-J. K. Liu, *IEEE Trans. Electron Devices, IEEE Trans. Electron Devices*, **64** (10), p. 4193 (2017).
22. L. F. Tiemeijer, R. M. T. Pijper, and E. v. d. Heijden, *IEEE Trans. Microwave Theory & Techniques,* **59** (3), p. 763 (2011).
23. J. Shi, A. Sidelnicov, Kok Wai J. Chew, Mei See Chin, C. Schippel, J.M.M. dos Santos, F. Schlaphof, L. Meinshausen, John R. Long, and D.L. Harame, *European Solid-State Device Research Conference*, p. 190 (2018).
24. P. Chevalier, G. Avenier, E. Canderle, A. Montagné, G. Ribes, and V.T. Vu, *IEEE IEEE Bipolar/BiCMOS Circuits and Technology Meeting (BCTM)*, p. 80 (2015).
25. Win Semiconductors, *European Microwave Week* (2018).

Advanced Transistors For High Frequency Applications

B. Parvais[a,b], U. Peralagu[a], A. Alian[a], A.Vais[a], L. Witters[a], Y. Mols[a], A. Walke[a], M. Ingels[a], H. Yu[a], V. Putcha[a], A. Khaled[a], R. Rodriguez[a], A. Sibaja-Hernandez[a], S. Yadav[a], R. ElKashlan[a,b], M. Baryshnikova[a], G. Mannaert[a], R. Alcotte[a], E. Simoen[a], S. E. Zhao[c], B. De Jaeger[a], D. M. Fleetwood[c], R. Langer[a], M. Zhao[a], P. Wambacq[a,b], B. Kunert[a], N. Waldron[a], and N. Collaert[a]

[a] imec, Kapeldreef 75, 3001 Leuven, Belgium
[b] Vrije Universiteit Brussels, Pleinlaan, 1050 Brussels, Belgium
[c] Vanderbilt University, Nashville, USA
correspondent author email: parvais@imec.be

Downscaling CMOS technology has allowed the integration of high-speed transceivers on silicon chips, but high-power amplifiers rely on III-V technologies to deliver the power and efficiency levels required by modern radios. In this work, we explore two routes to enable the fabrication of compound semiconductor devices on a large-scale manufacturable Si platform. In the first route, we report on Al(Ga,In)N HEMTs, MISHEMTs and MOSFETs integrated on 200 mm Si wafers using Au-free processing in standard Si CMOS tools. In the second route, we demonstrate GaAs/InGaP HBTs grown on a 300 mm Si substrate using Nano-Ridge Engineering (NRE) combined with aspect ratio trapping (ART). We provide insight on the potential of these new technologies for the design of advanced front-end modules, including performance trade-offs, modelling and reliability challenges.

Introduction

The advent of 5G will not only bring great new opportunities but also new challenges for the technologies enabling this next generation of mobile communications. The expectation for 5G is that it will enable extreme mobile broadband with peak data rates up to 10 Gbps and even higher (to address, e.g., the expected explosion in video streaming demand over the coming years), massive machine-to-machine communication to support the Internet-of-Things (IoT) platform and critical machine communication (e.g., autonomous driving) with ultra-high reliability and latencies below 1ms.

To meet this increasing demand for higher data rates, there is a push towards higher operating frequencies, moving from the congested sub 6 GHz bands to mm-wave, opening new research opportunities in the fields of devices, circuits, and systems. In the front-end of mm-wave transceiver systems, power amplifiers (PAs) are certainly one of the most important elements as their characteristics can limit the overall system performance (1,2). In particular, the success of the next mobile radio generations will rely on the achievable power efficiency and form factor. The limited area available for the radio module in mobile devices restricts the number of antenna paths available for beamforming. This, in turn, requires each PA to deliver a large output power (P_{out}), as

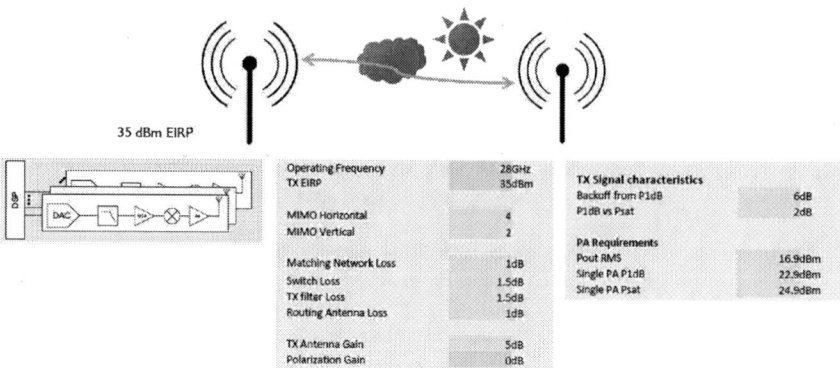

Figure 1 Example of a 28GHz link budget for mobile communication considering 8 antenna paths.

illustrated in the link budget in Figure 1. In order to ensure decent battery operation time, the optimization of the transmitter towards high power added efficiency (PAE) is crucial.

The choice of the appropriate technology should therefore be guided by its capability to realize power efficient transceivers. Interesting trends can be observed from the *PAE* vs P_{sat} data of various power amplifier technologies in the above 40 GHz regime (Figure 2) (3). Among the traditional high-frequency device technologies (GaAs, InP, SiGe, and GaN), III-V HBTs and HEMTs are the only technologies that have demonstrated good power performance across the full mm-wave spectrum. At medium power levels (20-30 dBm), these devices hold a significant power added efficiency (PAE) advantage over advanced CMOS and are competitive to GaN HEMTs, which are the preferred option for delivering very large power levels. These excellent performances are due to the intrinsic material properties (high electron sheet density, electron mobility and velocity, and a large band gap), leading to high values for the Johnson figure-of-merit. GaN has the extra advantage of having a similar thermal conductivity as Si, which is better than the GaAs one. The higher power handling capabilities of these compound semiconductor technologies, as opposed to CMOS and SiGe, are advantageous for mm-wave operation, where a higher P_{out} and PAE translate to a smaller number of elements required to drive the antennas and more energy-efficient systems with smaller form factor can be enabled.

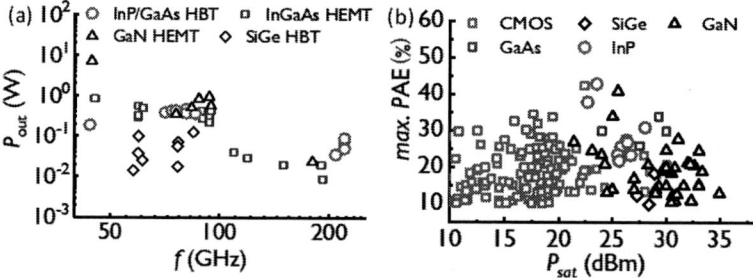

Figure 2 (a) Output power vs. operating frequency for on chip power amplifiers and transistors in the mm-wave regime. III-V HBTs hold clear advantage for the beyond-5G era (>40 GHz). (b) Maximum PAE for various power amplifier technologies in the above 40 GHz regime (data taken from (3)). For medium power range applications (20-30 dBm, e.g. for user held devices), III-V HBTs and HEMTs can be the technologies of choice.

However, current GaAs and GaN technologies are still limited to small size, expensive, non-Si substrates and use older generation processing (4). Migrating to a 200- or 300-mm Si platform and, manufacturing devices using standardized CMOS fab tools are critical steps toward the uptake of GaN and GaAs devices for RF and mm-wave applications (5).

In this paper, we describe technological solutions for the large-scale manufacturing of compound semiconductors, compatible with CMOS processing. First, the integration and optimization of Al(Ga,In)N devices on 200 mm Si substrate using Au-free, CMOS-compatible processing is discussed. Second, we demonstrate GaAs/InGaP HBTs grown on a 300 mm Si substrate using nano-ridge engineering.

GaN transistors

Various types of AlGaN/GaN and AlInN/GaN devices (MOSFET, MISHEMT and HEMT) fabricated on 200 mm Si substrates are studied in this section (6).

Substrate

Compared to SiC, using Si as the starting substrate material is known to degrade the RF losses, which are critical for the integration of switches and passive components in the RF front-end module. The RF loss is dominated by the interface loss due to the p-type conductive channel at the AlN/Si interface, which is induced by the thermal diffusion of Al during the high-temperature growth. Although a low growth temperature of the AlN nucleation layer can limit the RF loss in the AlN/Si template, it results in a low crystalline quality of AlN for practical use. The optimization of the epitaxy conditions is therefore essential to obtain a good balance between the crystalline quality, morphological quality, and RF performance (7). Furthermore, the resistivity of the starting Si should be high in order to limit the RF losses.

The epitaxial structure used in the present study is grown by MOCVD on high resistivity, 200 mm Si (111) substrates, and is composed of a proprietary GaN/AlGaN buffer structure (8), a GaN channel, a 1 nm AlN spacer, a 19 nm $Al_{0.25}Ga_{0.75}N$ barrier and a 5 nm in-situ grown Si_3N_4 cap for surface passivation. By optimizing the growth conditions, we obtained crack-free epilayers of high, uniform quality and low wafer warp of $< \pm 40$ µm, which is compliant with CMOS fab processing.

Using a high resistivity Si substrates, combined with careful buffer design, C-doping process at a lower growth temperature and strict control of the thermal budget during the growth of the active region, enabled a very low RF transmission loss (evaluated by the attenuation constant from coplanar waveguide structures) of 0.15 dB/mm at 20 GHz, as shown in Figure 3. By careful optimization of the process conditions, it is possible to obtain effective substrate resistivities above 2kΩcm, which is somewhat lower than what trap-rich SOI can provide today, but higher than typical HR-SOI substrates.

It is essential to reduce buffer thickness for GaN-on-Si RF devices in order to ensure thermal and mechanical stability of the GaN-on-Si wafers during fabrication, thus improving the yield and reducing the fabrication costs. However, reducing buffer thickness comes at the cost of a decrease in the vertical breakdown voltage ($V_{BD,vertical}$) (9). The buffer voltage required to attain a leakage current of 1µA/mm² at T=25°C and 10µA/mm² at T=150°C is defined as $V_{BD,vertical}$, and is observed to be higher than 100V for buffers with thickness up to 1.3mm (Figure 4).

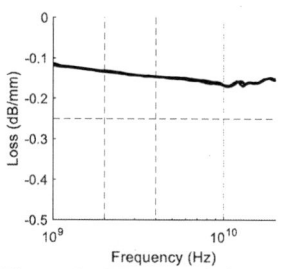

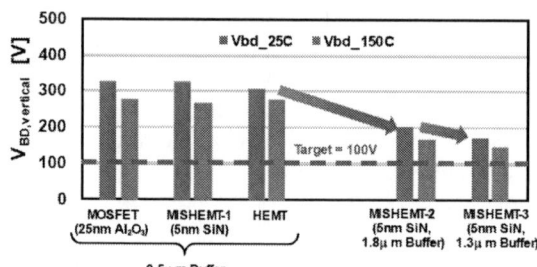

Figure 3 RF transmission loss (evaluated by the attenuation constant) of the device buffer up to 20 GHz as measured from 2 mm long coplanar waveguides.

Figure 4 Median buffer breakdown voltage ($V_{BD}, vertical$) >100V reduces for decreasing buffer thickness.

Devices fabrication

Five AlGaN/GaN device splits were realized based on differences in gate processing for HEMT, MISHEMT and MOSFET. fabrication (Figure 5). The devices were processed up to Metal 1 using the Au-free, gate-first process described in (Figure 6) (6). One limitation of 200 mm processing is the lack of a lift-off or dry etch process for Ni, necessitating the use of TiN as the gate metal which has a lower Schottky barrier height. Different gate stacks are further considered to minimize the gate resistance. The minimal gate length featured in the present study is approximately 300nm.

An AlInN barrier has also been explored since it is expected to bring further benefits (6). Owing to the high spontaneous polarization, a similar or larger electron density can be obtained for a thinner barrier. This enhances the lateral and vertical scalability of the device, which is important for f_T/f_{max}. In addition, the ability to lattice match InAlN to GaN obviates the internal stresses that could otherwise be a reliability concern. An improved contact resistivity ($R_C<0.2\Omega cm$) was observed for the AlInN barrier wafer, which could arise from differences in surface treatment before ohmic metallization and reduction of Schottky barrier height from the presence of In.

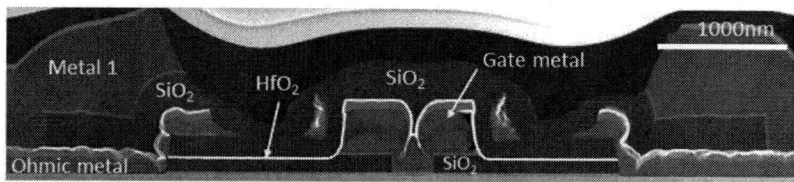

Figure 5 TEM photograph of a fully processed device.

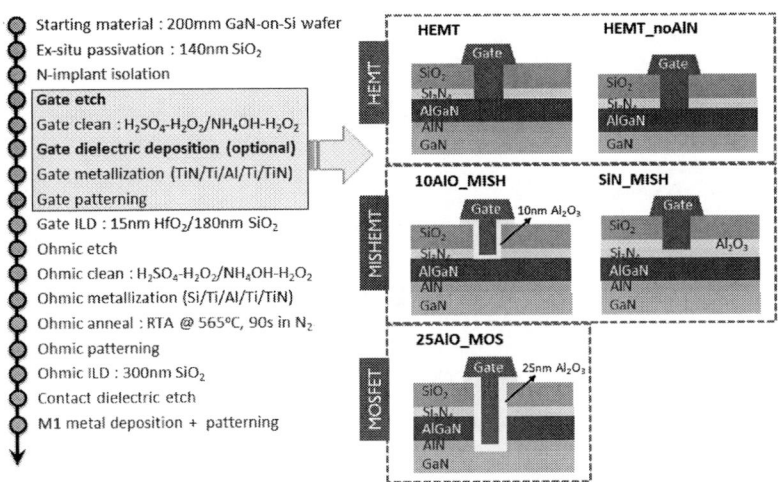

Figure 6 Gate-first process flow for the fabrication of GaN devices. Five device splits were realized based on differences in gate processing for HEMT, MISHEMT and MOSFET.

DC performance

The I-V characteristics in Figure 7 enables one to understand the specificities of the considered device architectures. The HEMT has the lowest I_{ON}/I_{OFF} ratio of 10^3 and improves to 10^4 in the absence of AlN, but at the cost of on-state performance. The MOSFET has the highest I_{ON}/I_{OFF} ratio of 10^8, although suffering from an additional parasitic component contributing to the higher R_{EXT}, which needs further study. Short channel effects result in a negative V_T at $L_G = 300$ nm for the MOSFET. The HEMTs/MISHEMTs show a trend of decreasing V_T for increasing gate to channel distance, with a lower V_T being needed to deplete the channel when AlN is present due to the higher N_S. The reduction of oxide/barrier thicknesses is key therefore to suppress SCE with further L_G scaling. The MISHEMT Si_3N_4 cap is only 2 nm after processing explaining the relatively higher I_G.

The interplay between gate capacitance (C_G) and mobility lead to the observed trends for extrinsic transconductance (G_M) (Figure 8). For HEMT splits, a higher G_M is obtained with AlN, which apart from improving carrier confinement, screens the channel electrons from the barrier, thereby reducing alloy scattering, and improving μ_{FE}. The MISHEMTs show the highest μ_{FE} (>2000 cm^2/V.s). Despite this, the reduction in C_G from the oxide outweighs the mobility benefit, resulting in a lower G_M. The proximity of the oxide to the channel results in increased scattering, thus reducing the mobility of the MOSFET. Combined with the higher R_{EXT} and large C_G from the thick oxide, this gives rise to the low G_M. This highlights that the MISHEMT, with an appropriately scaled oxide/barrier, which does not impact on leakage, can achieve a high mobility, while keeping a large C_G.

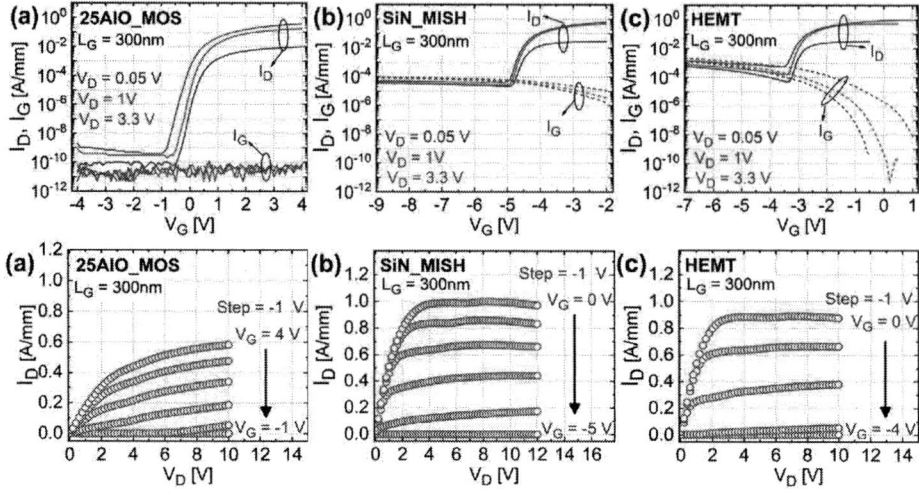

Figure 7 Transfer and output characteristics of (a) MOSFET, (b) MISHEMT and (c) HEMT.

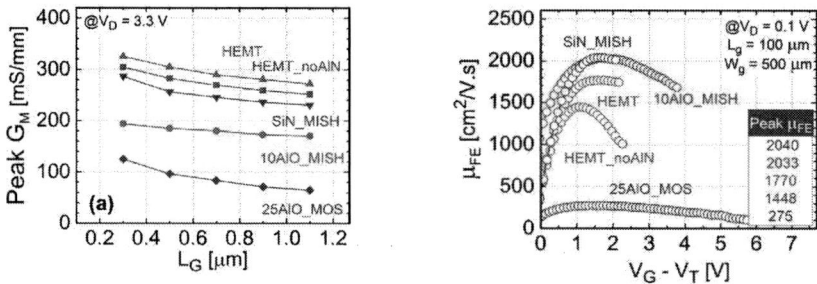

Figure 8 (a) Peak G_M versus L_G for AlGaN barrier devices. (b) Channel field effect mobility (μ_{FE}) versus gate overdrive (V_G-V_T) for AlGaN barrier devices.

RF performance

The peak cut-off frequencies obtained from S-parameters measurements on non-optimized transistors are f_T/f_{max} = 46/38 GHz at L_G = 300 nm, W_f=100um. The f_T is aligned to the state-of-the art at this dimension and is suited for operation in the sub 6 GHz bands. RF performance for mm-wave operation would greatly benefit from gate length downscaling (Figure 9).

In contrast to f_T, the cut-off frequency of the unilateral power gain, f_{max}, does depend on the finger width, as a result of its dependence to gate resistance:

$$f_{max} \approx \frac{f_t}{2\sqrt{g_d\left(R_g + R_s + R_i\right) + 2\pi f_t R_g C_{gd}}}$$

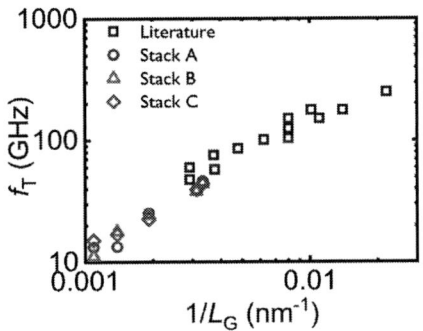

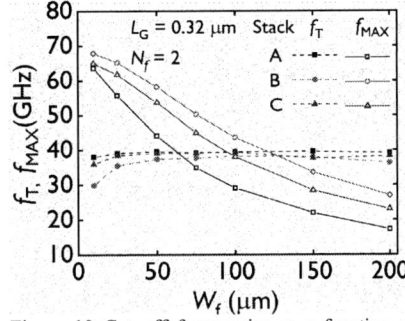

Figure 9 Gate scaling of f_T and benchmark with literature.

Figure 10 Cut off frequencies as a function of finger width for 3 different gate stacks.

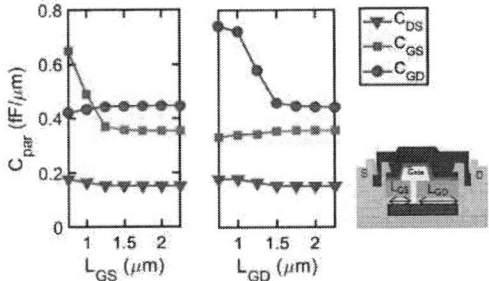

Figure 11 Parasitic capacitance extracted around 10 GHz as a function of the gate to source/drain distance.

Therefore, gate stack optimization enables an improvement of f_{MAX} up to ~50% using Ti-free AlCu based gate-metal stack (Figure 10) (10). In order to optimize the device for high frequency, the gate-to-source spacing must be minimized, to keep the source access resistance low. On the other hand, the reduction of gate to S/D spacing leads to an increase of the parasitic capacitance (Figure 11), which can deteriorate the overall RF performance if not properly optimized.

Variability and Reliability

Low frequency noise, measured on a HEMT, MISHEMT and MOSFET (Figure 12), is observed to follow the carrier number fluctuation model trend. This indicates that charge trapping in oxide defects close to the oxide/semiconductor interface is one of the major sources of the observed 1/f noise (11). Another possible source could be bulk-semiconductor defects. It is evident that the noise power spectral density of the drain current (S_{Id}) is the lowest for the MISHEMT. This is expected as the GaN channel is separated from the oxide by the AlGaN barrier. This increases the effective tunneling distance for charge carriers in the channel into the oxide, thus considerably reducing the oxide-trapping component as a dominant source of 1/f noise. Conversely, MOSFETs have the highest S_{Id}, as the GaN channel forms a direct interface with the gate oxide. Therefore, the oxide-trapping component becomes one of the dominant sources of 1/f noise. While HEMT devices lack a gate-oxide, they suffer from high gate-leakage, which results in higher S_{Id}.

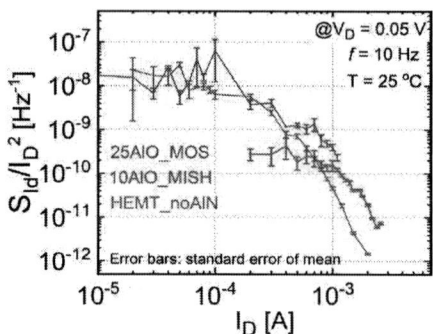

Figure 12 Noise power spectral density of the drain current (S_{Id}) for MOSFET, MISHEMT and HEMT. The error bars indicate the standard error of the mean (standard deviation/√no. of devices).

The critical DC-reliability metrics of future scaled GaN-on-Si RF devices are listed in Table 2. A thorough analysis of the listed DC-reliability parameters was presented in (9). The issues 1-2 pertain to the GaN buffer, while the issues 3-9 are related to the various aspects of the GaN RF device.

The MOSHEMT and MISHEMT devices were observed to be the most robust device types for scaled GaN-RF applications. This was attributed to the presence of the AlGaN barrier layer, which helps in alleviating some of the degradation modes. The absence of a barrier layer in MOSFETs results in poorer device performance due to the proximity of the gate-stack to the channel and the direct impact of oxide degradation on channel properties. The gate-leakage in HEMTs is inherently high due to the absence of a dielectric layer, which leads to early onset of buffer and lateral breakdown, increased buffer dispersion and HCI/SHE, thus critically limiting the device lifetime. A degradation-map based DC-reliability assessment was explored in (9) for the GaN RF devices and it was shown that further development of the degradation map acceleration model was required to assess the device long term reliability.

We note that while the interface trap distribution (Issues 5 and 9 in Table 1) at the AlGaN/oxide interface ($D_{it,ch/ox}$), and that of the GaN/AlGaN interface ($D_{it,ch}$) may not play a direct role in affecting the DC-reliability of the device, they are important device parameters which influence the dependence of oxide field on the applied gate-bias, thus affecting gate-stack reliability.

Table 1 Various DC reliability metrics and their impact on different devices.

Issue	Reliability Metric	MOSFET	MOSHEMT	MISHEMT	HEMT
1	$V_{BD,vertical}$				No barrier
2	Buffer dispersion		Least vertical field		Highest $I_{leakage}$
3	$V_{BD,lateral}$	Highest leakage		Full barrier = low E_{ox}	
4	TDDB lifetime	No barrier	Higher t_{ox},k		
5	D_{it} (channel/oxide)	Direct contact	Optimized barrier/oxide		
6	1/f noise PSD	Direct oxide interface		Most isolation	
7	ΔN_{eff} @operating E_{ox}	Unfavorable defect energies	Favorable defect dist.		
8	BTI/HCI/SHE lifetime				Highest leakage
9	σD_{it} (GaN/Barrier)	No Barrier			Direct impact on 2DEG

☐ Least impacted   ☐ Moderately impacted   ☐ Most impacted

GaAs HBT

We demonstrated recently a GaAs/InGaP HBT deposited on 300 mm Si substrates by Nano-Ridge Engineering (NRE) (12). This approach is further discussed in this section.

Device fabrication

NRE is based on selective area growth (SAG), that allows an easier co-integration of III/V devices with silicon, compared to the growth of a continuous hetero-film on blanket Si substrates (13). NRE leverages the aspect ratio trapping (ART) technique to ensure an efficient defect confinement in narrow oxide trenches. A large volume of III-V material is nevertheless needed to obtain practical HBT. A second oxide was therefore added to define the NR width and reduces sidewall deposition in the active hetero-layer. The shape of the nano-ridge is designed for optimal device functionalities. In particular, the region containing the sub-collector was grown so that it broadens quickly to touch the sidewalls of that oxide before the deposition of the remaining hetero-layers.

The process steps to obtain GaAs/InGaP based HBTs fabricated on GaAs/InGaP nano-ridges (InGaP is used as the emitter material), as well as sketch of the devices and resulting TEM picture is depicted in Figure 13.

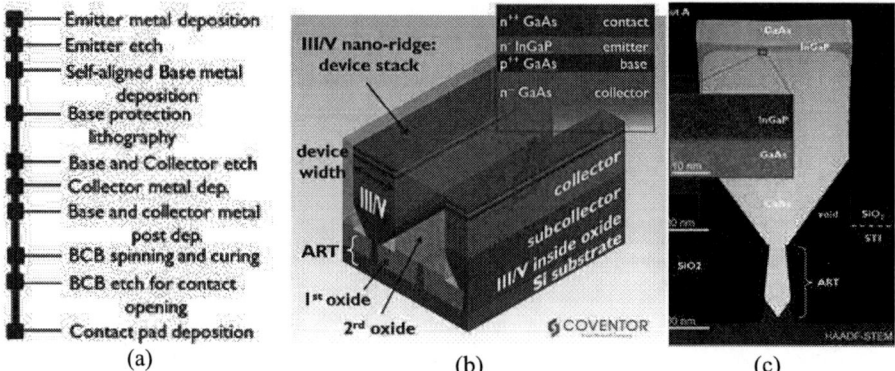

Figure 13 (a) Process flow. (b) Sketch of the NR HBT based on GaAs in a double-oxide template. (c) HAADF-STEM of GaAs/InGaP HBT stack after epitaxial deposition on a 300 mm (100) Silicon substrate along and across a nano-ridge.

Device performance

The measured device performance of the NR devices are comparable to those of a reference HBT structure deposited pseudomorphically on a 2 inch GaAs substrate under comparable MOVPE conditions (Figure 14), demonstrating the huge potential of this novel approach for the co-integration of III-V devices on Si wafers.

A large perimeter or nano-ridge sidewall leakage was nevertheless observed under forwards bias conditions. Perimeter plays an important role in the nano-ridge HBTs because of the large perimeter to area ratios inherent to these devices. This was also confirmed with TCAD simulations where interface traps at the nano-ridge sidewall and passivation interface were considered. As can be observed in Figure 15, the surface recombination in the junction space-charge region leads to a larger current compared to the junction diffusion current (15).

The small-signal performance of the devices in Figure 16 indicates a f_T value ~10GHz for this first, non-optimized demonstrator. Device RF performance can be boosted by both layout and process optimization. For instance, use of a high resistivity substrate will mitigate the impact of substrate parasitic on RF performance. The collector current density can be increased with further optimizing the collector contact process and using epitaxial regrowth for emitter, base, and collector regions. Emitter width scaling and employment of double contacts for base-collector electrodes will improve the high-frequency characteristics (especially f_{max}). Lastly, implementation of InP-based HBTs will lead to significant enhancement of cut-off frequencies.

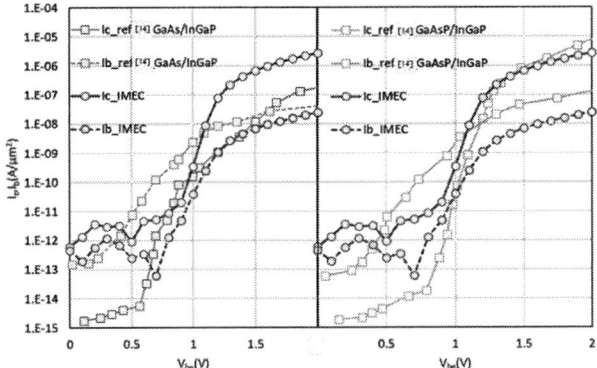

Figure 14 A comparison of electrical characterization (Gummel plot) of the devices fabricated in this work with that of ref. (14).

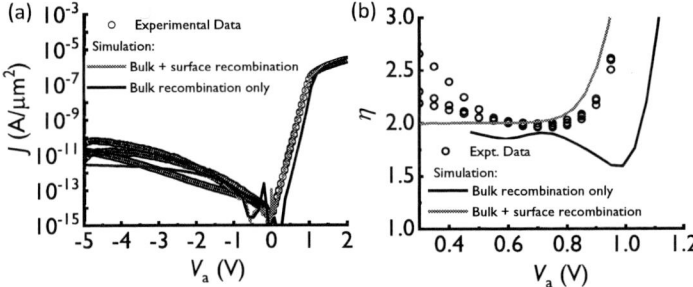

Figure 15 Impact of interface traps at GaAs/passivation on base-collector diode characteristics. Recombination in the junction at nanoridge sidewall explains the on-state behavior of the BC diodes.

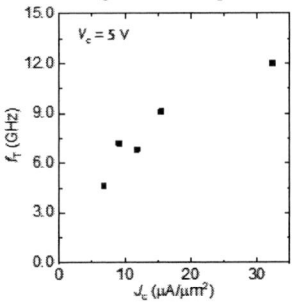

Figure 16 Current-gain cut-off frequency for devices with differnt emitter area.

Conclusions

High-power amplifiers are critical in modern radio systems and III-V and III-N technologies and enable delivering the required power and efficiency levels. Nanoridge engineering showed promising results for large scale Si-compatible integration of mid-power amplifiers, as a first step towards enabling InP-based devices. GaN MISHEMTs showed competitive performance for the integration of high power amplifiers in a CMOS compatible platform.

References

1. P.M. Asbeck *et al.*, *IEEE Trans. Microwave Theory and Tech.*, p. 3099, **67**(7) (2019).
2. Sherif Shakib *et al.*, *IEEE Communications Magazine*, p. 98, **57**(1), (2019)
3. Wang *et al.*, Power Amplifiers Performance Survey 2000-Present,[Online]. Available: https://gems.ece.gatech.edu/PA_survey.html .
4. K. Yuk et al., Proc *MWSCAS, p. 803* (2017).
5. HW. Then *et al, Proc IEEE Int. Electr. Dev. Meeting* (2019).
6. U. Peralagu *et al.*, *Proc IEEE Int. Electr. Dev. Meeting* (2019).
7. S. Chang *et al.*, *Semiconductor Science and Technology*, accepted for publication, available online (2020).
8. M. Zhao, 'Method for forming a semiconductor structure for a gallium nitride channel device', EP3486939A1 (2019).
9. V. Putcha *et al.*, accepted for publication in *Int. Rel. Phys. Symp.* (2020).
10. R. ElKashlan, *et al.*, submitted to *European Microwave Week* (2020).
11. D. M. Fleetwood, *IEEE Trans. Nucl. Sci.*, p. 1462, **62**(4) (2015).
12. A. Vais *et al.*, *Proc IEEE Int. Electr. Dev. Meeting* (2019).
13. B. Kunert *et al.*, *Compound Semiconductor*, **24** (5) (2018).
14. C. Heidelberger and E.A. Fitzgerald, *J. of Appl. Phys.* 123, 161532 (2018).
15. S. Yadav *et al.*, submitted to *European Microwave Week* (2020).

Junctionless Device Cross-section: A Key Aspect for Overcoming Boltzmann Tyranny

Abhinav Kranti and Manish Gupta

Low Power Nanoelectronics Research Group, Discipline of Electrical Engineering,
Indian Institute of Technology Indore, India
Email: akranti@iiti.ac.in

In this work, we report on the significance of device cross-section
of a 3-dimensional Junctionless transistor (JLT) to facilitate a
sharp rise in drain current with a low value (< 10 mV/decade) of
subthreshold swing (SS). The analysis shows that apart from the
usual device parameters such as gate length and drain voltage, the
cross-section of a tri-gate JLT offers an additional degree of
freedom to tune the extent of impact ionization in JLT. Results
demonstrate that tri-gate JLT designed with lower aspect ratio (<
0.5) and wider film is most appropriate to overcome Boltzmann
tyranny. The work highlights the usefulness of cross-sectional area
for achieving sharp rise in drain current in JLT.

Introduction

The switching functionality of a transistor can be characterized by the current transition
from off-to-on state. In general, the drain current transition is defined by Subthreshold
swing (SS), also referred as inverse subthreshold slope, which is essentially limited by V_t
$\times \ln(10)$, where V_t is the thermal voltage (1). At room temperature (300 K), the minimum
value of SS is equal to 60 mV/decade (1). However, when impact ionization is
prevalent, the minimum value of SS can be reduced to a value lower than 60
mV/decade (2). This is particularly interesting as SS < 60 mV/decade translates into a
higher on-to-off current ratio (~10^5) at relatively lower applied voltages. In order to
overcome the fundamental SS limit of 60 mV/decade, various device architectures
such as impact ionization MOS transistor (I-MOS) (2-3) and tunneling based field
effect transistor (TFET) (4) have been proposed and analyzed. While TFETs are
potential contenders to yield sub-60 mV/decade current transition at a lower value of
drain bias (V_{ds}), the switching action critically depends on the source/drain doping profile
and gate-to-source overlap (4). However, this improvement in switching is achieved at
the expense of parasitic capacitance. Although lower values of SS can be achieved using
I-MOS transistor, their potential is limited by requirement of higher supply voltage to
trigger impact ionization. Therefore, architectures facilitating impact ionization at
comparatively lower drain biases should be considered.

A gated resistor, also known as junctionless transistor (JLT), has a much simpler
architecture and has a heavily doped semiconductor film with only one type (n or p) of
impurity atoms (5). JLT has shown immunity towards surface roughness scattering due to
its unique bulk conduction mechanism (5-14). Moreover, when sufficiently higher drain
bias is applied, a JLT (even with only one type of dopant) exhibits impact ionization and
is able to achieve a sharp rise in drain current with a lower than 60 mV/decade

subthreshold swing at a drain bias lower than that required for an inversion mode MOSFET and I-MOS (15-22). The primary reason for the enhanced degree of impact ionization is higher current density (J) exhibited by heavily doped JLT (18) which lead to the generation of electrons and holes in the semiconductor film. In addition, authors in (21-22) have experimentally demonstrated that JLT is immune towards hot electron degradation and show a marginal change in the steep SS with successive measurement.

The downscaling of the transistor as predicted by Moore's law is foreseen to continue with multiple gate architecture due to better gate controllability and lower degree of short channel effects (1). While offering improved gate controllability, a tri-gate JLT provides an additional degree of freedom to tune the impact ionization phenomenon for sub-60 mV/decade current transition through its cross-sectional area. As JLT is a heavily doped device, the cross-sectional area governs the concentration of carriers, and thus, current density in the channel available for ionization triggered steep SS. Therefore, this work investigates the influence of cross-sectional area on steep switching in tri-gate JLT. The cross-sectional dependence is considered by evaluating the aspect ratio (AR) in a tri-gate JLT, and results showcase the new possibilities through the optimization of cross-section of a 3-dimensional JLT for attaining a sharp enhancement of drain current from off-to-on state.

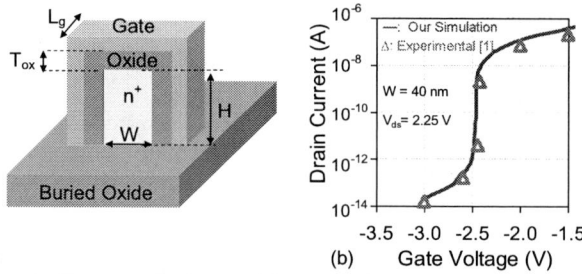

(a)   (b)   Gate Voltage (V)

Figure 1. (a) Schematic diagram of tri-gate silicon (Si) junctionless transistor (JLT), and (b) Comparison of simulation result with published experimental data (15) of a tri-gate JLT.

Simulations

Tri-gate Silicon (Si) nMOS JLT, shown in figure 1a, has been analyzed using numerical simulations through Silvaco TCAD software (23). The cross-section of a tri-gate JLT was defined through the fin-width (W) and fin-height (H). As the focus of the work is on cross-section, JLT had a relatively thick oxide (T_{ox}) of 10 nm and the semiconductor film was uniformly doped with 10^{19} cm^{-3} donor atoms. In order to capture the essential physics of JLT, mobility model with field and doping dependence along with surface roughness, acoustic phonons, and optical intervalley phonon scattering mechanisms. Since impact ionization is integral to the analysis, modules corresponding to bipolar phenomenon, impact ionization and bandgap narrowing due to heavy doping were included (23). As shown in figure 1b, the models used in the simulations were able to capture the sharp increase in current and compared well with the published experimental result (15) for tri-gate JLT. As the focus of the work was cross-section optimization, the values of W was varied for a fixed H which resulted in various values (0.2 to 1) of aspect ratio ($AR = H/W$).

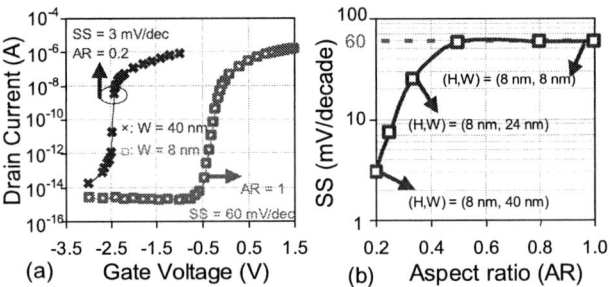

Figure 2. (a) Drain current (I_{ds}) - gate voltage (V_{gs}) characteristics of tri-gate Si JLT with varying AR exhibiting different SS values, and (b) Variation of SS with AR.

Results and Discussion

In order to understand the influence of cross-sectional area on the performance of tri-gate JLT, figure 2a compares the drain current (I_{ds})–gate voltage (V_{gs}) characteristics of JLT for two different combinations of H and W. Result depicts that JLT with a thicker film while maintaining a low AR of 0.2 achieves a steep rise in I_{ds} with $SS < 5$ mV/decade. Further, successive increase in AR to 1 degrades SS to a more conventional value of 60 mV/decade (figure 2b) which can also be seen from the more gradual I_{ds} transition (figure 2a). The reduction in SS from 60 mV/decade to a very steep value (< 5 mV/decade) is due to the impact ionization, which results in the generation of excess carriers in the film. The electrons generated due to impact ionization support the already existing electron concentration while the impact generated holes accumulate in the lower potential region. Hole concentration, which was minimal in the heavily doped n-type film, increased significantly due to impact ionization. Thus, the cumulative effect of sufficiently higher hole and electron concentration results in a sharp transition of current. Further, as bias is the same for both structures, the results showcase the importance of cross-section for achieving a sharp increase in drain current.

In general, MOSFETs are designed with narrow films (or higher AR) to minimize the short channel effects as the two vertical gates (apart from the top gate) can control the channel potential. However, in contrast, results shown in figure 2a-b reveal that JLT with a lower AR is beneficial to reduce SS. As the channel region in JLT is heavily doped, the number of carriers available for triggering impact ionization depends on the cross-section area of JLT. Therefore, a device with lower AR i.e. 0.2 offers higher number of carriers as compared to the architecture with $AR = 1$. This can be confirmed from figure 3a where the maximum electron concentration ($n_{e,max}$) in the off-state ($V_{gs} = -3$ V) is shown to increase as AR reduces. The higher value of $n_{e,max}$ indicates a strong degree of impact ionization (through higher current density) and leads to the generation of more number of holes as shown in figure 3b. The hole concentration was evaluated at the gated surface at an overdrive of 10 mV i.e. $V_{gs} = V_{th} + 10$ mV, where V_{th} is the threshold voltage. Figure 3b shows that maximum hole concentration in the film reduces from $\sim 10^{20}$ cm^{-3} (at $AR = 0.2$) to $\sim 8 \times 10^{18}$ cm^{-3} (at $AR = 1$) which indicates the dominant role of JLT cross-section for triggering impact ionization.

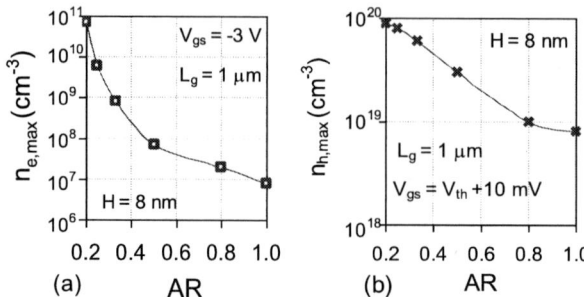

(a)   AR   (b)   AR

Figure 3. Variation of maximum (a) electron ($n_{e,max}$) and (b) hole ($n_{h,max}$) concentration with AR. The electron concertation was extracted at the location of conduction channel at $V_{gs} = -3$ V (off-state), whereas the impact generated hole concertation was evaluated at a gate overdrive of 10 mV (on-state) at the region of minimum potential.

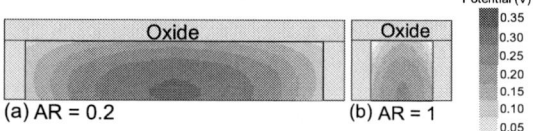

(a) AR = 0.2   (b) AR = 1

Figure 4. (a) 2D contour plot at $V_{gs} = -3$ V (off-state) showing the potential distribution in tri-gate JLT designed with an (a) $AR = 0.2$ and (b) $AR = 1$.

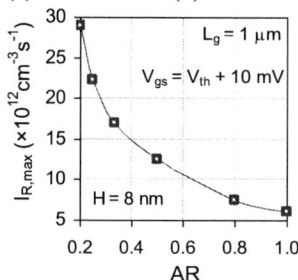

Figure 5. Dependence of peak magnitude of ionization rate ($I_{R,max}$) with varying AR at $V_{gs} = V_{th}+10$ mV.

In order to further comprehend the occurrence of steep SS in JLT with lower AR, figure 4*a-b* compares the electrostatic potential in two different topologies i.e. $AR = 0.2$ and 1. Figure 5 shows the variation in peak magnitude of ionization rate, extracted at $V_{gs} = V_{th}+10$ mV, with AR. While comparing both the topologies, it is clear that JLT with $AR = 0.2$ shows significantly higher I_R as a wider area is available in the bulk of the device over which impact ionization occurs. As availability of excess carriers and wider area are key contributors to impact ionization, a device with lower AR is more conducive for impact ionization and associated steep switching. Also, the extent of depletion (in off-state) in a tri-gate JLT with lower AR is predominantly governed by the top gate (quasi planar), whereas in a JLT architecture with $AR = 1$, all three gates control the depletion of carriers which is obviously enhanced as compared to the device with lower AR value. The variation in I_R is consistent with behavior of $n_{e,max}$ as it reduces from ~3×10^{13} cm^{-3}s^{-1} ($AR = 0.2$) to 6×10^{12} ($AR = 1$).

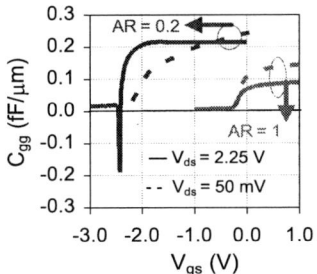

Figure 6. Variation of total gate capacitance (C_{gg}) with gate bias (V_{gs}) in tri-gate Si JL transistor for $AR = 0.2$ and 1 for low (50 mV) and high (2.25 V) values of drain bias (V_{ds}).

Figure 6 compares the total gate capacitance (C_{gg}) as a function of gate voltage in a JLT with $AR = 1$ and 0.2. While the JLT with $AR = 1$ shows a gradual increase in C_{gg} with increasing gate voltage, a sharp negative peak of C_{gg} is observed at the threshold voltage in JLT with $AR = 0.2$ at higher V_{ds} values. The occurrence of negative peak of C_{gg} is a characteristic feature of impact ionization induced sub-60 mV/decade switching. As shown in figure 3b, the maximum hole concentration in a JLT with $AR = 0.2$ reaches high value $\sim 10^{20}$ cm^{-3}. Thus, the higher hole concentration generated due to impact ionization 'selectively' changes the heavily doped n-type region near the gate (lower potential region) into a pseudo p-region which results in negative values of C_{gg} at the threshold voltage. Capacitance characteristics at lower V_{ds} (50 mV) do not show negative C_{gg} values due to the absence of impact ionization for all AR values.

Conclusion

The significance of device cross-section for triggering impact ionization and achieving sub-60 mV/decade off-to-on current transition has been presented. Even though depleted, a quasi-planar structure with very low aspect ratio (~ 0.2) is likely to support higher electron concentration in the off-state as compared to that with a relatively higher aspect ratio (~ 1). This essentially translates into a higher current density which can lead to an enhanced degree of impact ionization and associated steep subthreshold swing in tri-gate junctionless transistors.

Acknowledgment

This work is supported by the Science and Engineering Research Board (SERB), Department of Science and Technology (DST), Government of India, under Grant CRG/2019/002937.

References

1. I. Ferain, C.A. Colinge, and J.-P. Colinge, *Nature*, **479**, 310-316 (2011).
2. K. Gopalakrishnan, P.B. Griffin and J.D. Plummer, *IEEE Trans. Electron Devices*, **52**, 69–76 (2005).
3. E.-H. Toh, G.H. Wang, L. Chan, G. Samudra and Y.-C. Yeo, *Semiconductor Science and Technology*, **23**, 015012 (2007).

4. C.D. Llorente, J.-P. Colinge, S. Martinie, S. Cristoloveanu, J. Wan, C. Le Royer, G. Ghibaudo, M.Vinet, *Solid-State Electronics*, **159**, 26-37 (2019).
5. J.-P. Colinge, C.-W. Lee, A. Afzalian, N. Akhavan, R. Yan, I. Ferain, P. Razavi, B. O'Neill, A. Blake, M. White, A.-M. Kelleher, B. McCarthy and R. Murphy, *Nature Nanotechnology*, **5**, 225-229 (2010).
6. J. Jiang, J. Sun, W. Dou and Q. Wan, *IEEE Electron Device Letters*, **33**, 65-67 (2012).
7. G. Zhang, Q. Wan, J. Sun, G. Wu and L. Zhu, *IEEE Electron Device Letters*, **34**, 265-267 (2013).
8. J. Guo, J. Liu, B. Yang, G. Zhan, L. Tang, H. Tian, X. Kang, H. Peng, X. Chen and C. Yang, *IEEE Electron Device Letters*, **37**, 908-910 (2015).
9. Y. Song, P.K. Mohseni, S.H. Kim, J.C. Shin, T. Ishihara, I. Adesida and X. Li, *IEEE Electron Device Letters*, **37**, 970-973 (2016).
10. J.-Y. Lin, M.P.V. Kumar and T.-S. Chao, *IEEE Electron Device Letters*, **39**, 8-11 (2018).
11. J.-W. Han, D.-I. Moon and M. Meyyappan, *IEEE Electron Device Letters*, **38**, 1156-1158 (2018).
12. M.P.V. Kumar, J.-Y. Lin, K.-H. Kao and T.-S. Chao, *IEEE Trans. on Electron Devices*, **65**, 3535-3542 (2018).
13. M. Seo, M.-H. Kang, S.-B. Jeon, H. Bae, J. Hur, B.C. Jang, S. Yun, S. Cho, W.-K. Kim, M.-S. Kim, K.-M. Hwang, S. Hong, S.-Y. Choi and Y.-K. Choi, *IEEE Electron Device Letters*, **39**, 1445-1448 (2018).
14. A. Vandooren, J. Franco, B. Parvais, Z. Wu, L. Witters, A. Walke, W. Li, L. Peng, V. Deshpande, F.M. Bufler, N. Rassoul, G. Hellings, G. Jamieson, F. Inoue, G. Verbinnen, K. Devriendt, L. Teugels, N. Heylen, E. Vecchio, T. Zheng, E. Rosseel, W. Vanherle, A. Hikavyy, B.T. Chan, R. Ritzenthaler, G. Besnard, W. Schwarzenbach, G. Gaudin, I. Radu, B.-Y. Nguyen, N. Waldron, V. De Heyn, D. Mocuta and N. Collaert, *IEEE Trans. Electron Devices*, **65**, 5165-5171 (2018).
15. C.-W. Lee, A. Afzalian, N. Akhavan, R. Yan, I. Ferain, and J.-P. Colinge, *Applied Physics Letters*, **94**, 053511 (2009).
16. S.M. Lee and J.T. Park, *IEEE Trans. Electron Devices*, **60**, 3856-3861 (2013).
17. R. Yu, A.N. Nazarov, V.S. Lysenko, S. Das, I. Ferain, P. Razavi, M. Shayesteh, A. Kranti, R. Duffy and J.-P. Colinge, *Solid-State Electronics*, **90**, 28-33 (2013).
18. M.S. Parihar, D. Ghosh, and A. Kranti, *Journal of Applied Physics*, **113**, 184503 (2013).
19. M. Gupta and A. Kranti, *IEEE Trans. Electron Devices*, **64**, 2061-2066 (2017).
20. M. Gupta and A. Kranti, *IEEE Trans. Electron Devices*, **65**, 2406-2412 (2018).
21. S.M. Kim, C.G. Yu, W.-J. Cho and J.T. Park, *IEEE Trans. Electron Devices*, **64**, 2526–2532 (2017).
22. S.-Y. Kim, B.-H. Lee, J. Hur, J.-Y. Park, S.-B. Jeon, S.-W. Lee and Y.-K. Choi, *IEEE Electron Device Letters*, **39**, 4-7 (2018).
23. ATLAS Users Manual, Silvaco, 2018.

Discussion on the Figures of Merit of Identified Traps Located in the Si Film: Surface versus Volume Trap Densities

B. Cretu[a], B. Nafaa[a], E. Simoen[b,c], G. Hellings[b], D. Linten[b], and C. Claeys[d]

[a] Normandie Univ, UNICAEN, ENSICAEN, CNRS, GREYC, 14000 Caen
[b] Imec, Kapeldreef 75, B-3001 Leuven, Belgium
[c] Solid-State Physics Department, Ghent University, 9000 Gent, Belgium
[d] Dept. Electrical Engineering, KU Leuven, Kasteelpark Arenberg 10, B-3001 Leuven, Belgium

> The aim of this work is a discussion on the figures of merit of identified traps located in the depletion zone (Si film) of advanced MOSFET devices. Two methodologies to estimate the volume trap densities are investigated, one using the relationship between the surface trap density and volume trap density and a second one based on the temperature evolution at fixed frequency of the generation-recombination plateau level associated to the same trap. By comparing the volume trap densities estimated using these two methods, the results are not agreeing with each other, suggesting that these methods can no longer be used with accuracy in multi-gate devices. Moreover, they may lead in certain cases to results physically not correct. Even about of the volume defects, the linear evolution between the plateau and the characteristic frequency of the generation-recombination contributions associated to the same trap give us the surface trap density without any additional assumption.

Introduction

Low frequency noise is a powerful non-destructive electrical diagnostic tools for predicting the semiconductor device quality (1,2). In particular low frequency noise spectroscopy may give information on processing-induced defects in scaled MOSFET structures, allowing the study of deep-level traps in the gate stack or in the semiconductor material, whatever are the device dimensions and architecture (3,4).

In this work, the estimation of the density of traps located in the depletion area (Si film) of devices processed using different state-of-the-art MOSFET architectures is assessed through different methodologies (5,6). The first method to estimate the volume traps density uses the relationship between the surface trap density and volume trap density. Another method for volume traps density extraction is related to the temperature evolution at fixed frequency of the generation-recombination plateau level associated with the same trap, it is shown that the obtained volume densities depend on the chosen fixed frequency of the generation-recombination level at which the estimation is made. The results obtained using these two methods are not in agreement with each other suggesting that they may be inappropriate to estimate the volume traps density in multi-gate devices. Contrary, from the linear behavior of the generation-recombination plateau versus the time constant (both associated to the same trap) the effective trap density can be estimated, and this with no additional assumptions. Moreover, the importance of plotting this generation-recombination plateau versus the time constant is related to the fact that it may be considered as a supplementary confirmation of the trap identification that has been carried out.

In the second section, details on the investigated devices and the experimental set-up are given. In the third section, useful considerations and equations related to generation-recombination noise and the methodologies to estimate the density of traps located in the Si film are reviewed. Finally, in the fourth section, critical discussion on the different methods to estimate the volume and surface density of traps located in the Si film is given before wrapping up.

Experimental

The discussion on the figures of merit (surface and volume trap densities) is performed for identified traps located in the depletion region of different multi-gate transistors (e.g. FinFETs, UTBOX, GAA NW FETs) processed at imec (Leuven, Belgium) in fully depleted (FD) SOI (Silicon on Insulator) technologies. More technological details, e.g. the channel gate length and width ratio, the thickness of the non-intentionally doped Si film, the gate stack composition and the equivalent oxide thickness may be found in (7,8). The low frequency noise measurements were made directly at wafer-level using a Lakeshore TTP4 prober. The home-made noise measurement set-up allows to bias the devices by choosing the polarization voltages using standard supply voltages. The current noise at the output of the devices is converted into a voltage noise using an I to V converter. A low noise voltage amplifier and a HP3562A spectral analyzer are used to obtain the noise power spectral density. The device input-referred noise power spectral density is calculated by dividing the measured output voltage noise power spectral density by the square of the measured voltage gain between the gate and the output. More details on the experimental set-up are provided in (9).

The Fermi level changes with the applied gate bias. Maximum generation recombination noise is created where the Fermi level and the traps level cross in the bandgap. Since the energy level of a point defect located in the depletion region is discrete and unique, and when the applied gate bias change, the Fermi level scans the same trap, but for increasing depth in the depletion zone (4,6). The characteristic time constant of the generation recombination noise associated with this trap will not change with gate bias variation but should only vary with temperature (3,4,6). A constant drain current polarization is necessary to keep a quasi-constant Fermi level over all the targeted temperature range. Performing low frequency noise spectroscopy measurements require to maintain a constant drain current polarization by adjusting the gate voltage at a fixed drain bias. The methodology to estimate the noise parameters is described in (10).

Methodology - useful equations

In linear operation, the gate voltage spectral density of the generation − recombination due to traps located in the depletion region of the transistor is expressed as (3 ,6):

$$S_{Vg\_Lor}(f) = \frac{q^2 N_{eff} \tau_i}{WLC_{ox}^2} \frac{1}{1 + (2\pi f \tau_i)^2}$$

[1]

where q is the absolute electron charge, N_{eff} is the surface trap density, W and L are the effective channel width and length, and C_{ox} is the gate capacitance per unit of area.
Each generation-recombination contribution is characterized by a plateau level A_i and a time constant τ_i (4,5,6):

$$A_i = \frac{q^2 N_{eff}}{WLC_{ox}^2} \tau_i = \frac{q^2 BW_d N_T}{WLC_{ox}^2} \tau_i \qquad [2]$$

where W_d is the silicon depletion depth (e.g. equal to the Si film thickness T_{Si} for UTBOX devices or $W_{Fin}/2$ for FinFETs, where W_{Fin} is the fin thickness) and N_T is the volume trap density.
From the slope of the linear behavior which should exist between A_i and τ_i (associated to the same trap) extracted for all temperatures where the identified trap is active, the surface trap density N_{eff} can be directly estimated.
The B coefficient which permits to estimate the volume trap density from the slope of the A_i and τ_i (associated to the same trap) parameters is defined as 1/3 [5,6]. This is the usual method to estimate the volume density of the identified traps in conventional planar MOSFET technologies (named Method 1).
A second method to estimate the volume trap density consists to use the maximum of the measured $S_{Vg\_Lor}(f_0,T)$ dependence with temperature (named Method 2). Indeed, the $S_{Vg\_Lor}(f_0,T)$ of generation-recombination noise associated to the same trap is proportional with $\tau_i(T)/\{1 + [2\pi f_0 \tau_i(T)]^2\}$. For a given frequency f_0, if $2\pi f_0 \tau_i(T) \gg 1$, $S_{Vg\_Lor}(f_0,T) \propto \tau_i(T)^{-1}$, and $S_{Vg\_Lor}(f_0,T)$ increases with increasing temperature because τ_i decreases. If $2\pi f_0 \tau_i(T) \ll 1$, then $S_{Vg\_Lor}(f_0,T) \propto \tau_i(T)$ and $S_{Vg\_Lor}(f_0,T)$ decreases with increasing temperature, as explained in detail in (6).

Results and discussion

A trap related to hydrogen V_2H was identified in standard <100> and rotated <110> UTBOX n-type transistors (11). From the slope of A_i and τ_i (Eq. 2) a surface trap density of $1.2 \cdot 10^9$ cm^{-2} was obtained for the standard device and of $8.1 \cdot 10^9$ cm^{-2} for the rotated one. Considering B as 1/3 (5,6), and taking into account that the Si film thickness (T_{Si}) is about 16 nm after device processing, this leads to a volume trap density N_T of $2.2 \cdot 10^{15}$ cm^{-3} for the standard device and of $1.5 \cdot 10^{16}$ cm^{-3} for the rotated one.
The evolution of the $S_{Vg\_Lor}(f_0,T)$ with the temperature at fixed frequency is presented in Figure 1 for both devices. From the maximum of the bell-shaped behavior the volume trap density may be estimated. The results are summarized in Table 1.

TABLE I. Summary of estimated surface and volume V_2H traps densities for a UTBOX device.

Double gate device	N_{eff} (cm^{-2})	N_T (cm^{-3})		
		Method 1	Method 2	
standard channel UTBOX	$1.2 \cdot 10^9$	$2.2 \cdot 10^{15}$	$f_0 = 4$ Hz	$7 \cdot 10^{15}$
rotated channel UTBOX	$8.1 \cdot 10^9$	$1.5 \cdot 10^{16}$	$f_0 = 5$ Hz	$3 \cdot 10^{16}$
			$f_0 = 8$ Hz	$3.3 \cdot 10^{16}$
			$f_0 = 12$ Hz	$3.8 \cdot 10^{16}$

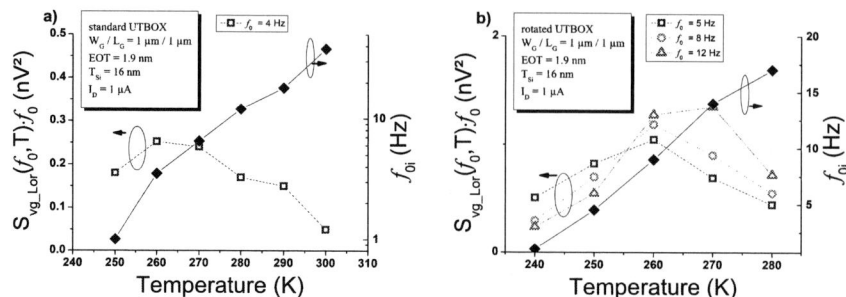

Figure 1. $S_{Vg\_Lor}(f_0,T) \cdot f_0$ versus temperature for the V_2H trap identified in (11); on the secondary Oy axis the characteristic frequency f_{0i} of the generation recombination noise is displayed in function of temperature.

As observed from Figure 1b, the maximum of the $S_{Vg\_Lor}(f_0,T)$ behavior is dependent on the fixed frequency that was considered. Moreover, regarding the results of Table I, Method 2 provides higher values compared to Method 1 for both standards and rotated devices.

Concerning the triple-gate devices (FinFETs), an example of the evolution of the $S_{vg\_Lor}(f_0,T) \cdot f_0$ in a temperature range for a trap most likely related to the C_iC_s complex is given in Figure 2.

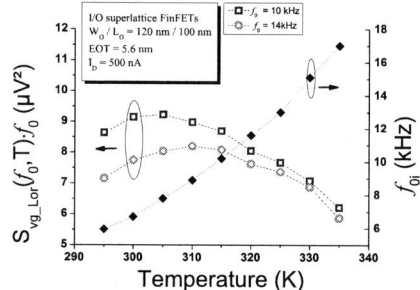

Figure 2. $S_{Vg\_Lor}(f_0,T) \cdot f_0$ versus temperature for the C_iC_s trap identified in (11); on the secondary Oy axis the characteristic frequency f_{0i} of the generation recombination noise is displayed in function of temperature

From the slope of A_i versus τ_i a surface trap density of this trap of $2.8 \cdot 10^{12} cm^{-2}$ was obtained (10).

Using Method 1 gives a volume trap density of about $1.7 \cdot 10^{19}$ cm^{-3}.

Using Method 2, volume trap densities of about $1.6 \cdot 10^{18}$ cm^{-3} at $f_0 = 10$ kHz and of about $1.4 \cdot 10^{18}$ cm^{-3} at $f_0 = 14$ kHz are obtained. It may be observed that the estimated volume trap density of this defect is about one decade lower than when using Method 1.

This trend was observed for all identified traps in the FinFET. The results are summarized in Table II.

Taking into account that the channels are non-intentionally doped (N_A of about 10^{15} - 10^{16} cm^{-3}), the obtained values of the volume trap densities in these multi-gate devices (FinFETs) may seems unphysical, whatever method (1 or 2) is employed. It should be noticed that when generation-recombination contributions of different traps have close characteristic time constants this may lead to an overestimation of the volume trap densities when using Method 2.

TABLE II. Summary of estimated surface and volume densities of identified traps for a FinFET device.

Triple gate device FinFET	N_{eff} (cm^{-2})	N_T (cm^{-3})		
		Method 1		Method 2
V_2H	$6.2 \cdot 10^{10}$	$3.7 \cdot 10^{17}$	$f_0 = 20$ Hz	$9.4 \cdot 10^{16}$
$V_2(0/-)$	$2.2 \cdot 10^{11}$	$1.32 \cdot 10^{18}$	$f_0 = 140$ Hz	$3.5 \cdot 10^{17}$
V-P	$8.5 \cdot 10^{11}$	$5.1 \cdot 10^{18}$	$f_0 = 1.2$ kHz	$1.3 \cdot 10^{18}$
C_iC_s	$2.8 \cdot 10^{12}$	$1.7 \cdot 10^{19}$	$f_0 = 10$ kHz	$1.6 \cdot 10^{18}$
			$f_0 = 14$ kHz	$1.4 \cdot 10^{18}$

A last example presented is for a GAA NW FETs with an identified V_2H trap (12). The results of the estimated surface and volume traps densities are summarized in Table III. The same trend as for FinFETs is observed: Method 2 gives lower volume trap densities compared to Method 1. One should note that the considered "rectangular" gate-all-around devices having a fin height and fin width equals to 10 nm, the depletion zone is taken as 5 nm.

TABLE III. Summary of estimated surface and volume V_2H trap densities for a GAA NW FET device.

Gate all around device (GAA NW FET)	N_{eff} (cm^{-2})	N_T (cm^{-3})		
		Method 1		Method 2
V_2H	$3 \cdot 10^9$	$1.8 \cdot 10^{16}$	$f_0 = 80$ Hz	$9 \cdot 10^{15}$

It can be observed that the estimation of volume traps densities using Method 1 and Method 2 does not match very well.

It is important to remind that Method 2 can be applied to estimate the density of the noisy centers for both "generation" and "trapping" noise, while the B = 1/3 approach is for "generation" noise (6). Furthermore, by comparing the volume trap densities obtained using Method 2 with the estimated values of the surface trap densities, one can estimate the experimental B coefficient, expressed as

$$B_{exp} = \frac{N_{eff}}{N_T W_d} \qquad [3]$$

The obtained values are summarized in Table IV. The fact that the obtained values of B_{exp} coefficient are lower (factor of 2 or 3) than the theoretical one (1/3) was already reported for UTBOX devices (11) and it was suggested that this trend may be linked to the fact that the theoretical B coefficient was theoretically evaluated for conventional planar transistor with one gate. However, for the GAA NW FET the B_{exp} is about 2 times higher than the theoretical value. In any case, the B_{exp} is lower than 1. Contrary, for FinFETs, the B_{exp} takes values higher than 1, which is unphysical.

TABLE IV. Determination of the experimental B (B_{exp}) (Note: are considered : for rotated UTBOX only the case of $f_0 = 8$ Hz, for the C_iC_s traps in FinFETs only the case of $f_0 = 10$ kHz)

device	standard UTBOX	rotated UTBOX	FinFET				GAA NW FET
trap	V_2H	V_2H	V_2H	$V_2(0/-)$	V-P	C_iC_s	V_2H
B_{exp}	0.1	0.15	1.32	1.25	1.3	3.5	0.66

This may suggests that for multi-gate devices, the methods permitting the calculation of the volume traps density developed for conventional planar transistors is no more accurate. The use of the volume trap density as figure of merit can be questioned.

As the surface trap density can be extracted directly from the slope of A_i versus τ_I (associated to the same trap) without any approximation, it is suggested here that it can be used as a figure of merit when comparing the density of traps located in the depletion region for transistors belonging to different technologies and architectures.

Conclusion

Disagreement between the obtained values of the volume traps densities when using different estimation methods is evidenced. Unphysical higher values of the estimated volume trap densities calculated using both methods is found for the FinFET case. In certain cases, unphysical values for the experimental B value are found. This suggests that the use of the volume density traps located in the depletion region as figure of merit for advanced multi-gates devices should be questioned. The effective trap densities estimation, without considering any additional hypothesis, could be used as a figure of merit even if the traps in the depletion region of the transistors are related to a volume phenomenon.

References

1. F. Scholz and J. Roach, *Solid-State Electron.*, **35**, 447 (1992).
2. D.C. Murray, A. Evans, and J.C. Carter, *IEEE Trans Electron Devices.*, **38**, 407 (1991).
3. V. Grassi, C.F. Colombo, and D.V. Camin, *IEEE Trans Electron Devices.*, **48**, 2899 (2001).
4. E. Simoen, B. Cretu, W. Fang, M. Aoulaiche, J.-M. Routoure, R. Carin, S.D. dos Santos, J. Luo, C. Zhao, J.A. Antonio Martino, and Cor Claeys, *Phys. Status Solidi C.*, **12**(3), 292 (2015).
5. L.D. Yau and C.-T. Sah, *IEEE Trans Electron Devices.*, **(16)**, 170 (1969)
6. N. Lukyanchikova, edited by A. Balandin American Scientific, Riverside, CA 2002, p. 201 (2002).
7. S.D. dos Santos, B. Cretu, V. Strobel, J.-M. Routoure, R. Carin, J.A. Martino, M. Aoulaiche, M. Jurczak, E. Simoen, and C. Claeys, *Solid-State Electron.*, **97**, 14 (2012).
8. G. Hellings, H. Mertens, A. Subirats, E. Simoen, T. Schram, L.-A. Ragnarsson, M. Simicic, S.-H. Chen, B. Parvais, D. Boudier, B. Cretu, J. Machillot, V. Peña, S. Sun, N. Yoshida, N. Kim, A. Mocuta, D. Linten, and N. Horiguchi. *in Tech. Dig. Symp. on VLSI Technology, The IEEE New York*, 85 (2018).
9. D. Boudier, B. Cretu, E. Simoen, R. Carin, A. Veloso, N. Collaert, and A. Thean. *Solid State Electron.*, **128**, 102 (2017).

10. D. Boudier, B. Cretu, E. Simoen, G. Hellings, T. Schram, H. Mertens, and D. Linten, *Solid State Electron.*, in press, (available online 2019).
11. B. Cretu, E. Simoen, J.-M. Routoure, R. Carin, M. Aoulaiche, and C. Claeys., *In Proceedings of ICNF'2015,* IEEE Xplore (2015)
12. A. Bordin, B. Cretu, R. Carin, E. Simoen, G. Hellings, D. Linten, C. Claeys. accepted to EuroSOI-ULIS'2020.

Impact of Gate Dielectric Material on Basic Parameters of MO(I)SHEMT Devices

P. G. D. Agopian[a,b], G. J. Carmo[a], J. A. Martino[b], E. Simoen[c], N. Collaert[c]

[a] UNESP, Sao Paulo State University, Sao Joao da Boa Vista, Brazil
[b] LSI/PSI/USP, University of Sao Paulo, Sao Paulo, Brazil
[c] imec, Leuven, Belgium

In this work, the behavior of MISHEMT devices with different gate materials is analyzed. Two-gate insulator materials (Al_2O_3 and SiN) were analyzed through the transfer characteristic, threshold voltage, hysteresis and transconductance. Although devices with SiN insulator present smaller hysteresis, better DIBL and it is nearest to a normally off devices, the leakage current showed to be much higher than for the Al_2O_3 counterpart. Besides the double conduction that occurs in SiN devices results in an anomalous behavior of transconductance and consequently an unexpected behavior of threshold voltage with temperature.

Introduction

Recently the high electron mobility transistors (HEMT) have drawn the attention of the scientific community for power electronic applications thanks to their wide bandgap and consequently high breakdown voltage (1-3) and good thermal conductivity.

Since GaN and AlGaN materials have different bandgaps, when these materials are put together, a region with a strong confinement of electrons (that can only move in two directions) is created. This region is called 2 Dimensional Electron Gas (2DEG), that acts as a conduction channel. The HEMT transistors has two modes of operation, normally ON and normally OFF. A critical problem with this device is its high gate leakage current, which is a roadblock for developing high power and low noise applications (1).

A new metal oxide/insulator semiconductor high electron mobility transistor (MOSHEMT) appears in order to minimize the leakage current as reported in (4). In this work, the basic parameters of MOSHEMT are evaluated experimentally for different gate insulators (10nm of Al_2O_3 and 5nm of SiNx).

Device Description

Figure 1 shows the cross-section view of the AlGaN / GaN MISHEMT device with different gate insulators, Al_2O_3 and SiN. Both structures consists of a layer of GaN grown on a 200mm Si substrate, followed by 15nm of AlGaN layer and the insulator covered by TiN gate material. Two different splits were evaluated: the first one with an insulator of 10nm of Al_2O_3 and the second with 5nm of SiN.

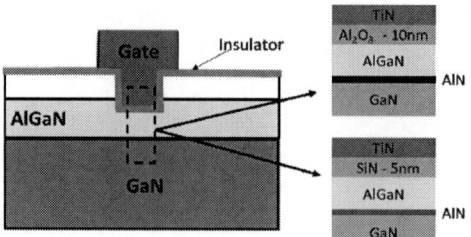

Figure 1: Schematic view of studied MISHEMTs

Results and Analysis

From the transfer curves presented in figure 2, one can notice a shift in the threshold voltage and different hysteresis behavior when comparing both insulators. The hysteresis analysis shows a better behavior for SiNx devices, which present negligible hysteresis values.

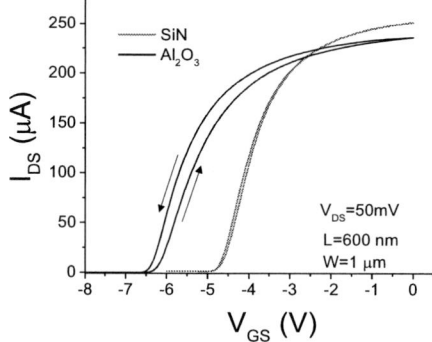

Figure 2: Experimental drain current as a function of gate voltage for devices fabricated with different gate dielectrics.

Figure 3 presents the hysteresis width extracted from devices with Al_2O_3 as a function of gate length for different drain bias. Although usually the Vth variation, in GaN HEMTs, is associated with an electron trapping or a degradation of the on-resistance of the device (3), which causes a positive Vth shift, in this case the opposite Vth variation was observed. This negative Vth shift may also be related with a variation of the on-resistance, which makes the device more distant from the desired normally-off behavior. As the drain voltage and consequently the drain current increase, the Vth shift becomes smaller reducing the hysteresis values, as can be seen in figure 3.

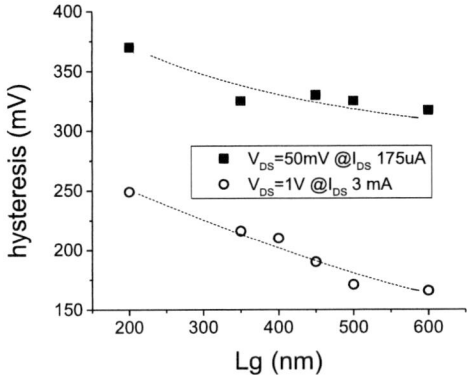

Figure 3: Hysteresis values as a function of channel length for MOSHEMTs with Al$_2$O$_3$ dielectric for 2 different drain bias.

Focusing on the threshold voltage variation between devices with different gate dielectrics, although both the gate insulator thicknesses and the permittivity impact on the threshold voltage (Vth) of the transistor in opposite direction (5), the dominant factor of the Vth shift, in this case, is a negative linear relation between Vth and the dielectric thicknesses. The MISHEMT with 5 nm of SiNx presents a Vth almost 2V higher than MOSHEMT with 10nm of Al$_2$O$_3$.

In addition to the better Vth value (closest to the normally-off device), SiNx dielectric transistors are less affected by the short channel effect (SCE) when compared to devices with Al$_2$O$_3$, as can be observed in figure 4.

From figure 4 it is possible to notice that reducing the channel length (Lg) from 600nm to 200nm, the threshold voltage for transistors with SiNx dielectric (figure 4B) drops 175mV while for devices with Al$_2$O$_3$ layer (figure 4A) this reduction reaches 450mV.

Drain Induced Barrier Lowering (DIBL) was calculated following the equation [1]

$$DIBL = \left(\frac{|Vth_{linear}-Vth_{sat}|}{V_{DS\,sat}-V_{DS\,linear}}\right) \qquad [1]$$

where $V_{DS\,sat}$ is the drain voltage at saturation region ($V_{DS\,sat}$=2V), $V_{DS\,linear}$ is the drain bias in triode regime ($V_{DS\,linear}$ = 50mV) and Vth$_{sat}$ and Vth$_{linear}$ are the threshold voltage at saturation and triode operation respectively.

Focusing on DIBL results, it is also observed that for both splits the DIBL starts to degrade for Lg = 400 nm, however the values and the degradation of DIBL are much higher for devices with Al$_2$O$_3$ than for devices with SiNx dielectric. This behavior can be related to the thinner dielectric and better coupling between the gate and the GaN layer.

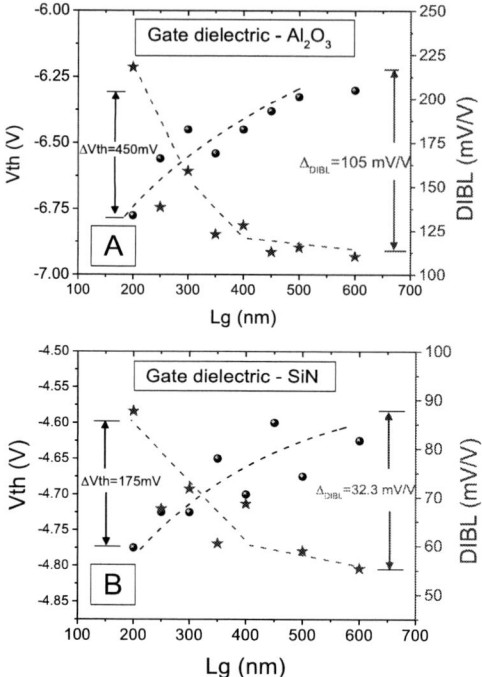

Figure 4: Threshold voltage and DIBL values for different channel lengths for MOSHEMTs with Al_2O_3 dielectric (A) and MISHEMTs with SiNx insulator (B).

Figure 5 presents the drain current as a function of gate voltage for temperatures ranging from 25°C to 150°C for MO(I)SHEMTs with both insulators, Al_2O_3 (A) and SiNx (B).

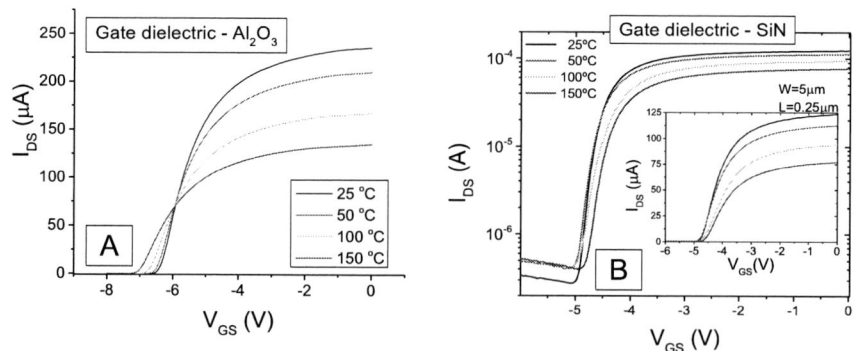

Figure 5: The transfer characteristic as a function of gate voltage for temperatures ranging from 25°C to 150°C for MISHEMT with Al_2O_3 dielectric (A) and with SiNx insulator (B).

When the MOSHEMTs were evaluated at high temperatures (from 25°C to 150°C), two different behaviors were obtained. The transfer characteristic of MOSHEMT with Al₂O₃ layer (figure 5A) presents an expected behavior, showing the zero temperature coefficient point (V_{ZTC}) due to the competition between the threshold voltage reduction and mobility degradation, while for the MISHEMTs with SiNx layer the V_{ZTC} was not found. In order to explain this unexpected behavior the threshold voltage and transconductance were evaluated (figures 6 and 7).

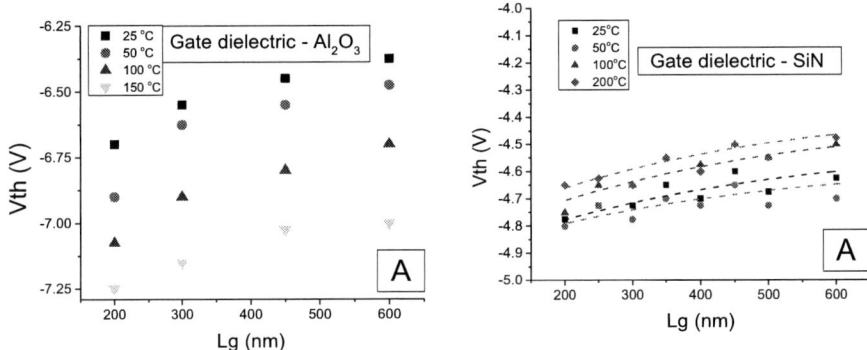

Figure6: Threshold voltage as a function of channel length for several temperatures for both types of devices: Al₂O₃ dielectric (A) and SiNx dielectric (B).

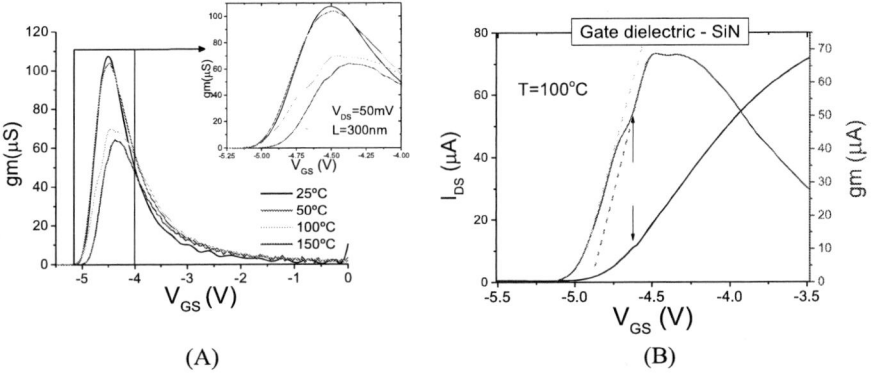

(A)   (B)

Figure 7: SiNx dielectric MISHEMT transconductance for device with W=5µm and L=300nm operating at different temperatures (A) and transconductance and drain current for MISHEMT operating at 100°C (B).

MOSHEMTs with Al₂O₃ gate present a reduction of the threshold voltage as the temperature increases (figure 6A) due to the reduction of the fermi potential, for all analyzed channel lengths. However, for SiNx MISHEMTs Vth is reduced when the temperature increases from 25°C to 50°C, but for higher temperatures Vth starts to increase again creating a rebound effect. The threshold voltage anomalous behavior can be explained by the double conduction of the device which leads us to extract an effective Vth.

This double conduction can only be observed when we evaluate the transconductance curve of the MISHEMTs (figure 7). It is possible to observe that for temperatures higher than 50°C the transconductance curve presents two different slopes before the maximum value, representing two different conduction channels. Figure 7B shows the transconductance and drain current as a function of gate voltage at 100°C and it is easy to see the two different slopes representing the two different threshold voltages, due to the turn on/off of the $SiN_x/AlGaN$ inversion layer. Looking for the transition point between both conductions, we can also observe a small peak in the drain current.

Conclusions

In this work, the influence of different types and thicknesses of the gate insulators on the basic parameters of the new MO(I)SHEMTs devices was evaluated. The split with gate insulator of SiN showed greater immunity to short channel effects, a less negative threshold voltage (behavior closer to the normally off device) and an excellent hysteresis analysis (hysteresis is practically suppressed). However, SiN devices also presents a double conduction (which did not occurs in Al_2O_3 devices), that results in an atypical behavior of threshold voltage with the temperature increase.

Acknowledgements

The authors thank CAPES, CNPq and FAPESP for the financial support.

References

(1) S. Gupta, S.N Mishra, K. Jena, , *International conference on Signal Processing, Communication, Power and Embedded System (SCOPES)*, 1777 (2016).
(2) J. Panda, K. Jena, R. Swain, T. R. Lenka, *Journal of Semiconductors*, **37**, 044003 (2016).
(3) F. Roccaforte, G. Greco, P. Fiorenza, F. Iucolano, *Materials*, **12**, 1599 (2019).
(4) R. Swain, K. Jena, T. R. Lenka, *Pramana- journal of Physics* **88**, 3 (2017).
(5) M. J. Dipsey, *Working towards a normally off GaN based MOSHEMT*, Master Thesis / Lehigh University, (2016).

On the Correlation Between Static and Low-Frequency Noise Parameters of Vertical Nanowire nMOSFETs

E. Simoen, A. Chasin, P. Matagne, E. Rosseel, A. Hikavyy, R. Loo P. Favia, H. Bender, E. Vancoille, and A. Veloso

Imec, Kapeldreef 75, B-3001 Leuven, Belgium

The low-frequency noise behavior of gate-all-around vertical nanowire silicon nMOSFETs is described and discussed with respect to the static device parameters, like the threshold voltage and the maximum transconductance. The spectra are dominated by 1/f noise at low frequencies, followed by white noise. The 1/f noise power spectral density at 10 Hz is correlated with the maximum transconductance and the threshold voltage. This is interpreted in terms of the diameter of the nanowires.

Introduction

The Gate-All-Around (GAA) Nanowire (NW) or nanosheet (NS) architecture provides superior short-channel effects control and is widely considered as one of the most promising candidates to replace finFETs in advanced logic technology nodes. Compared with their horizontal implementation, vertical NWFETs (1-2) offer some clear advantages, appearing promising devices to overcome some of the limitations encountered by conventional CMOS scaling as it reaches its physical limits and faces interconnect routing congestion. They have the potential to yield smaller circuits with reduced power consumption (3), allowing relaxed gate length without impacting the device footprint. While their fabrication is challenging, several integration schemes have been proposed for improved process control and device performance (1-2).

Besides a better control over the short-channel effects, it has been shown that NWs in many cases also yield an improvement in the low-frequency noise performance at lower drain current I_D (4-16). This has been assigned to the occurrence of volume inversion in the case of Inversion Mode (IM) transistors and to conduction in the core of the NWs for junctionless (JL) devices (13,17-21). Moreover, it has been shown that vertical GAA NWFETs may yield a lower noise Power Spectral Density (PSD) than horizontal architectures (14). On the other hand, at higher I_D, the LF noise PSD is typically enhanced by the impact of the rather high series or access resistance of NWFETs (12).

In this work, the static parameters of vertical NW junctionless (JL) nMOSFETs have been studied for devices with two different Replacement Metal Gate (RMG) cap layer widths (see schematics in Fig. 1), encapsulating the top part of the pillars during processing at the RMG module. It will be shown that this process parameter strongly impacts the static device parameters: the threshold voltage (V_T), the maximum transconductance (g_{mmax}) and the dynamic output resistance (r_{out}). Moreover, a correlation is shown with the input-referred voltage noise PSD (S_{VG}) at 10 Hz, which could be explained in terms of a process-induced variation in the diameter d_{NW} of the NWs.

Experimental Details

A detailed process description has been recently reported (2). In brief, a 55 nm in situ P-doped n^+ layer is grown by epitaxial deposition on a low-resistivity silicon substrate (source contact), followed by 90 nm undoped silicon and again 55 nm n^+ Si. The RMG stack consists of IL-SiO$_2$/HfO$_2$/TiN/W. An RMG cap consisting of SiO$_2$ is used to encapsulate the top of the pillars. Here, devices with two different cap widths and 100 pilars defined in arrays of 10 by 10 are compared. On-wafer DC and low-frequency noise measurements have been performed using a BT1500 Parameter Analyzer and an E4727A Advanced Low-Frequency Noise Analyzer from Keysight, respectively. Measurements have been performed in linear operation at a drain voltage V_{DS}=50 mV, while the gate voltage V_{GS} is stepped from weak to strong inversion. The S_{VG} is derived from the measured drain current noise PSD (S_I) by dividing through the measured transconductance squared ($g_m{}^2$) in each operation point. At least 10 devices per split have been evaluated to address the noise and static parameter variability.

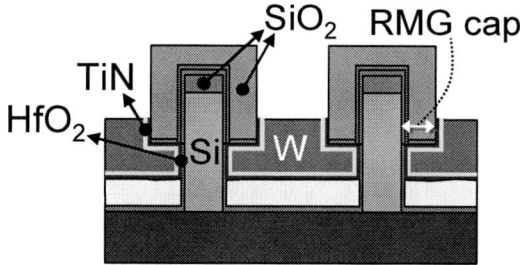

Figure 1. Schematic cross section of the vertical GAA NWFETs, with RMG SiO$_2$ cap.

Results and Discussion

Figure 2a and 2b compare the linear input I_D-V_{GS} characteristics of a set of narrow and wide cap vertical GAA nNWFETs. While there is obvious device-to-device variability, the wide-cap transistors (Fig. 2b) clearly exhibit a lower V_T than their narrow-cap counterparts in Fig. 2a. The transconductance (g_m) plots of Fig. 3 reveal another trend: the narrow cap devices with a high V_T (Fig. 3a) generally exhibit a lower maximum transconductance g_{mmax} and vice versa. This is even so within each flavour of devices in Figs 3a (narrow) and 3b (wide cap). Moreover, nFETs with a high(er) V_T and lower g_{mmax} also tend to exhibit a higher r_{out} (not shown).

The LF noise spectra, shown in Fig. 4 exhibit 1/f noise at low frequencies, followed by white noise. The 1/f noise becomes more pronounced for higher drain currents (from about 10 nA up to about 10 μA in Fig. 4), extending to the 1 kHz frequency range. The corresponding S_{VG} at 10 Hz is represented in Fig. 5. In the subthreshold regime, a decreasing trend of the 1/f noise PSD with gate voltage is observed, followed by a minimum around V_T. At the same time, devices with a lower threshold voltage also exhibit a lower S_{VG}. This is most clearly seen from comparing the data in Figs 5a (narrow cap) and 5b (wide cap) but can to some extent also be derived within each group of NWFETs.

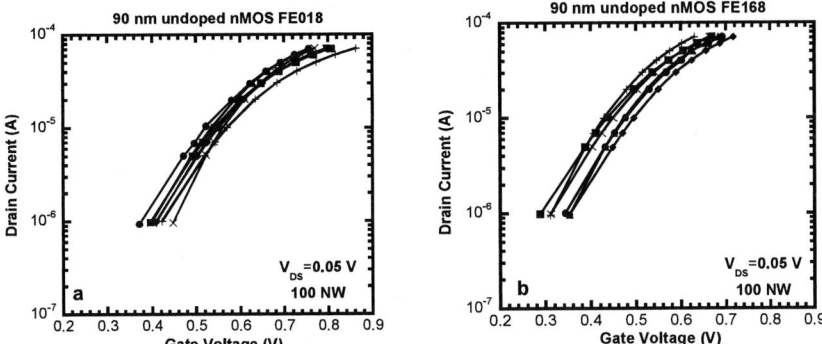

Figure 2. Input characteristics of n-type vertical GAA NWFETs with undoped channel and narrow (a) and wide (b) RMG cap.

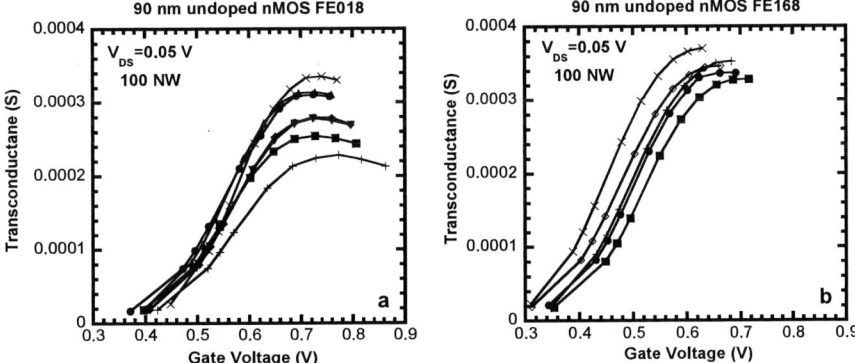

Figure 3. Transconductance in linear operation of n-type vertical GAA NWFETs with undoped channel and narrow (a) and wide (b) RMG cap.

The decrease of S_{VG} with V_{GS} reported here has been observed before for (vertical) NWFETs (5,9,15) and points to a so-called mobility fluctuations ($\Delta\mu$) origin of the 1/f noise. It is related to the fact that in the subthreshold regime, conduction occurs in the core of the undoped nanowire, away from the Si/SiO_2 interface. This suppresses carrier trapping by oxide traps and the associated number fluctuations 1/f noise. The increase at higher V_{GS} could then be ascribed to the impact of the series resistance, as observed for other NWFETs in the past (12). Overall, a minimum S_{VG} around V_T is found in Fig. 5, as a result of the two opposite mechanisms.

The observed correlation between the static and noise parameters is shown more explicitly in Fig. 6a and 6b, combining the narrow and wide-cap results of the foregoing Figs 2-5. Vertical GAA NWFETs with a low V_T correspond with a high g_{mmax} and with a smaller S_{VG} at 10 Hz and around V_T (minimum value in Fig. 5). Obviously, the dispersion in the noise values is significantly higher than in the static parameters.

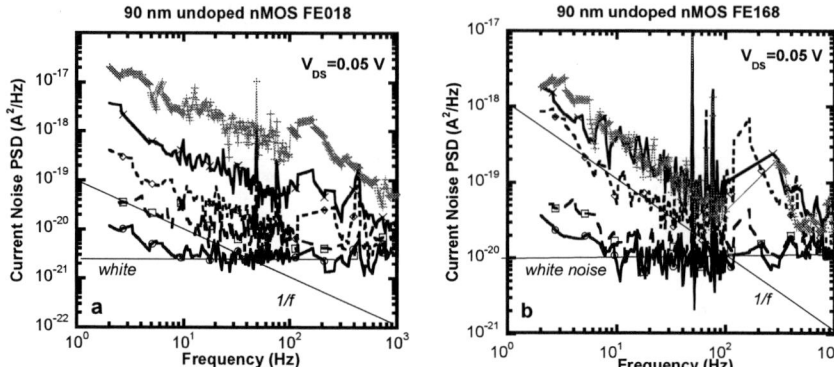

Figure 4. Low-frequency noise spectra of n-type vertical GAA NWFETs with undoped channel and narrow (a) and wide (b) RMG cap, in linear operation (V_{DS}=0.05 V and for different drain currents).

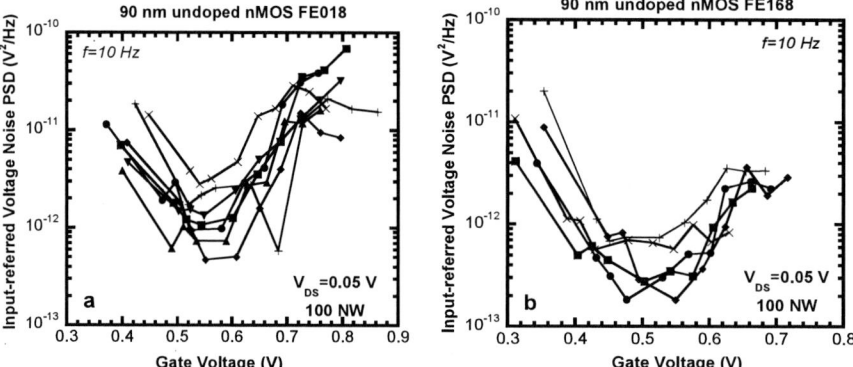

Figure 5. Input-referred voltage noise PSD at f=10 Hz and V_{DS}=50 mV of n-type vertical GAA NWFETs with undoped channel and narrow (a) and wide (b) RMG cap.

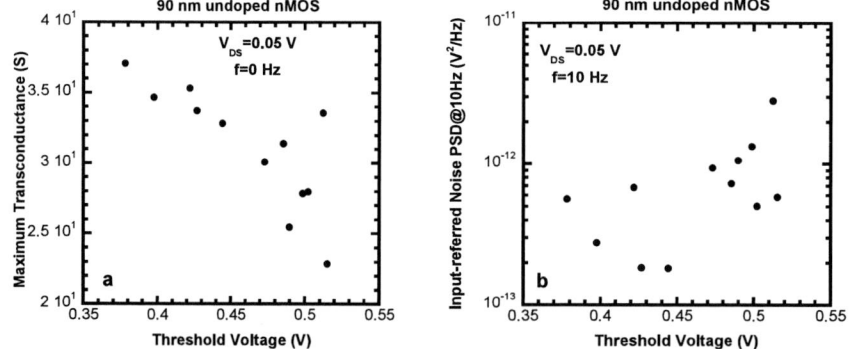

Figure 6. (a) Maximum transconductance in linear operation versus threshold voltage and (b) input-referred voltage noise PSD versus V_T for a set of 90 nm undoped n-channel vertical GAA NWFETs.

A possible explanation of the observed parameter variations should provide a consistent picture for the reported trends. For these specific devices, the NW doping variability can be ruled out as a possible explanation, as we are studying nominally undoped pillars. The difference in S_{VG} could suggest a difference in oxide/interface trap density, but this is not reflected in the subthreshold swing nor in the Positive Bias Temperature Instability (PBTI) behaviour, showing no impact (2). On the other hand, it has been reported before that the V_T of NW transistors increases for reduced d_{NW} (16). This could well explain also the lower g_{mmax} (scales with d_{NW}) and the higher S_{VG} (scales with $1/d_{NW}$). The different NW diameter could result from the different dummy dielectric removal for a different cap diameter (2). In addition, it has been reported before that the 1/f noise PSD of vertical pNWFETs reduces with d_{NW} (15), in line with our hypothesis.

Conclusions

The LF noise behavior of silicon vertical GAA nNWFETs has been reported, showing a 1/f noise that is dominated by mobility fluctuations in the subthreshold regime and by series resistance at high gate voltage overdrives. The minimum S_{VG} is shown to be correlated with the static device parameters: low PSD devices exhibit a lower V_T and a higher g_{mmax}. This could be interpreted in terms of the difference in nanowire diameter, associated with process variations. This is most clearly derived from comparing the performance parameters for the devices with a different RMG cap diameter: wide cap nNWFETs exhibit lower noise and V_T and a higher g_{mmax} compared with the narrow cap counterparts.

Acknowledgments

The devices have been processed in the frame of imec's Core Partner Program of Advanced Logic Devices.

References

1. A. Veloso *et al.*, in *Tech. Dig. of the Symposium on VLSI Technol.*, IEEE Xplore, p. 138 (2016).
2. A. Veloso *et al.*, in the *Proc. of IEDM19*, IEEE Xplore, p. 230 (2019).
3. T. Huyng-Bao *et al.*, in Proc. SPIE 2016, 978102 (2016).
4. Y. F. Lim, Y. Z. Xiong, N. Singh, R. Yang, Y. Jiang, D. S. H. Chan, W. Y. Loh, L. K. Bera, G. Q. Lo, N. Balasubramanian and D.-L. Kwong, *IEEE Electron. Device Lett.*, **27**, 765 (2006)
5. C. Wei, Y.-Z. Xiong, X. Zhou, N. Singh, S. C. Rustagi, G. Q. Lo and D.-L. Kwong, *IEEE Electron Device Lett.*, **30**, 668 (2009).
6. R.-H. Baek, C.-K. Baek, H.-S. Choi, J.-S. Lee, Y. Y. Yeoh, K. H. Yeo, D.-W. Kim, K. Kim, D. M. Kim and Y.-H. Jeong, *IEEE Trans. Nanotechnol.*, **10**, 417 (2011).
7. A. Veloso *et al.*, in *Proc. Symp. on VLSI Technol. Dig. of Tech. Papers*, IEEE Xplore, p. T138 (2015).

8. W. Fang, A. Veloso, E. Simoen, M.-J. Cho, N. Collaert, A. Thean, J. Luo, C. Zhao, T. Ye and C. Claeys, *IEEE Electron Device Lett.*, **37**, 363 (2016).
9. P. Singh, N. Singh, J. Miao, W.-T. Park and D. L. Kwong, *IEEE Electron Device Lett.*, **32**, 1752 (2011).
10. C.-H. Park, M.-D. Ko, K.-H. Kim, S.-H. Lee, J.-S. Yoon, J.-S. Lee and Y.-H. Jeong, *IEEE Electron Device Lett.*, **33**, 1538 (2012).
11. E. Simoen, A. Veloso, P. Matagne, N. Collaert and C. Claeys, *IEEE Trans. Electron Devices*, **65**, 1487 (2018).
12. J. Zhuge, R. Wang, R. Huang, Y. Tian, L. Zhang, D.-W. Kim, D. Park and Y. Wang, *IEEE Electron Device Lett.*, **30**, 57 (2009).
13. N. Clément, X. L. Han and G. Larrieu, *Appl. Phys. Lett.*, **103**, 263504 (2013).
14. T. Imamoto, Y. Ma, M. Muraguchi and T. Endoh, *Jpn. J. Appl. Phys.*, **54**, 04DC11 (2015).
15. C. Mukherjee, C. Maneux, J. Pezard and G. Larrieu, in *Proc. ESSDERC 2017*, IEEE Xplore, p. 34 (2017).
16. A. Veloso, P. Matagne, E. Simoen, B. Kaczer, G. Eneman, H. Mertens, D. Yakimets and B. Parvais, *J. Phys: Condens. Matter*, **30**, 384002 (2018).
17. W. Feng, R. Hettiarachchi, S. Sato, K. Kakushima, M. Niwa, H. Iwai, K. Yamada and K. Ohmori, *Jpn. J. Appl. Phys.*, **51**, 04DC06 (2012).
18. K. Ohmori, W. Feng, R. Hettiarachchi, Y. Lee, S. Sato, K. Kakushima, M. Sato, K. Fukuda, M. Niwa, K. Yamabe, K. Shiraishi, H. Iwai and K. Yamada, *ECS Trans.*, **45** (3), 437 (2012).
19. S.-H. Lee, C.-K. Baek, S. Park, D.-W. Kim, D. K. Sohn, J.-S. Lee, D. M. Kim and Y.-H. Jeong, *IEEE Electron Device Lett.*, **33**, 1348 (2012)
20. C. Liu, R. Wang, J. Zou, R. Huang, C. Fan, L. Zhang, J. Fan, Y. Ai and Y. Wang, in *IEDM11 Tech. Dig.*, IEEE Xplore, p. 521 (2011).
21. W. Feng, R. Hettiarachchi, Y. Lee, S. Sato, K. Kakushima, M. Sato, K. Fukuda, M. Niwa, K. Yamabe, K. Shiraishi, H. Iwai and K. Ohmori, in *IEDM11 Tech. Dig.*, IEEE Xplore, p. 630 (2011).

Intrinsic Voltage Gain of Stacked GAA Nanosheet MOSFETs Operating at High Temperatures

W.F. Perina [a], V.C.P. Silva [a], J.A. Martino[a], P.G.D. Agopian[a,b], E. Simoen[c], A. Veloso[c]

[a] LSI/PSI/USP, University of Sao Paulo, Sao Paulo, Brazil
[b] UNESP, Sao Paulo State University, Sao Joao da Boa Vista, Brazil
[c] Imec, Leuven, Belgium
e-mail: welder.perina@usp.br

In this work, the GAA silicon nanosheet MOSFETs basic parameters are evaluated for different channel lengths at high temperatures. The devices showed a subthreshold swing near the theoretical limit, low temperature variation on threshold voltage (dV_{TH}/dT = -0.4 mV/°C) and low drain induced barrier lowering (DIBL = 50 mV/V at 200°C), both for n-type device. The devices achieved an intrinsic voltage gain around 33 dB for the worst case (channel length of 28 nm), showing that this device is a promising technology for the 7 nm node of the MOS roadmap.

Introduction

In the semiconductor industry, the scaling down of devices has always been the driver in order to improve circuit area efficiency and high current drive, leading to the triple gate FinFET, which is the newest technology for high performance applications. However, it is believed that these devices cannot reach the sub 7 nm nodes and the GAA nanowires MOSFETs has emerged as one of the most promising candidates for this technology node, as it presents high density of integration and higher electrostatic control between gate and channel (1-2).

Devices Characteristics

The studied devices are GAA silicon nanosheet MOSFETs, fabricated at Imec, Leuven − Belgium, with two stacked channels with the vertical distance between the sheets of 7.5 nm, an EOT of 0.9 nm, a width of 15 nm and a channel length varying from 200nm down to 28nm. Figure 1 shows the TEM image of the device.

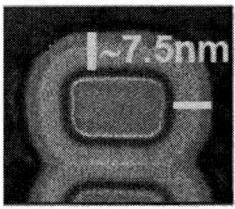

Figure 1. TEM image of the GAA silicon nanosheet MOSFETs.

Results

The threshold voltage (V_{TH}) as a function of temperature is presented in figure 2, in which it is possible to observe that, V_{TH} is reduced (in absolute value) for both n- and pMOS as temperature increases due to the fermi potential reduction. The V_{TH} variation rate for these devices shows to be better when compared to other technologies, presenting a variation around -0.64 mV/°C for the p-type, and -0.4 mV/°C for the n-type, while planar fully depleted SOI devices reach around -0.8 mV/°C (3), junctionless FinFETs can achieve around -0.94 mV/°C (4) and bulk nFinFETs present values around -0.67 mV/°C for narrow fins and -0.92 mV/°C for wide ones (5).

It is possible to see that the subthreshold swing (SS) values (figure 3) are very close to the theoretical limit for the MOS technology, except for 28nm channel length pMOS devices, indicating that pMOS is more susceptible to short channel effects (SCEs) than the nMOS counterpart.

In figure 4, the Drain Induced Barrier Lowering (DIBL) degrades when the temperature increases. DIBL degradation for both n- and pMOS with temperature are very similar, showing a variation rate of 0.11 mV/V.°C. It is also important to note that although DIBL is affected by the temperature, the obtained values (except for L=28nm) are very small reaching in the worst case 50mV/V at 200°C.

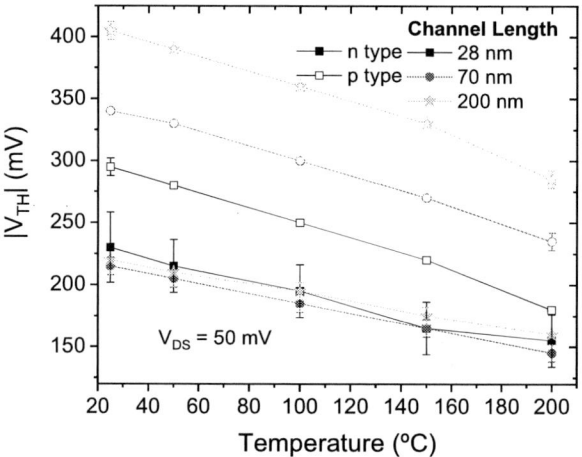

Figure 2. Threshold Voltage as a function of temperature for different transistors channel lengths. N-type transistors are represented by closed and P-type by open symbols.

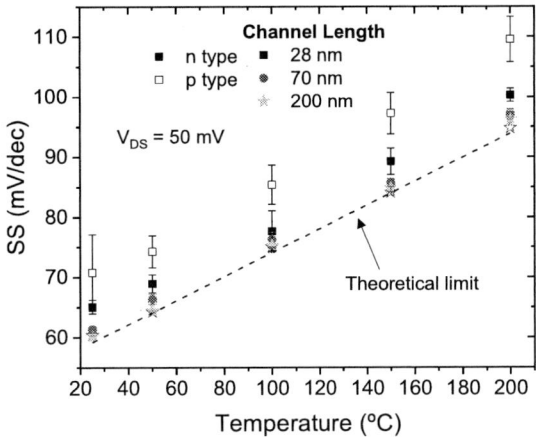

Figure 3. Subthreshold Swing as a function of temperature for P and N-type transistors with different channel lengths.

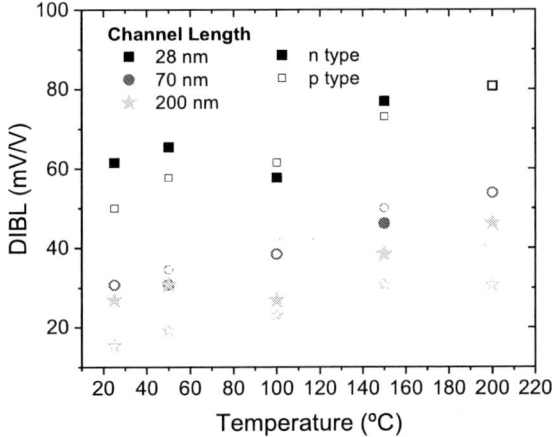

Figure 4. Drain Induced Barrier Lowering (DIBL) as a function of temperature for channel length varying from 200nm down to 28nm.

The transconductance (gm_{sat}) and output conductance (gd) on saturation region of n-type and p-type, showed in figure 5, decreases with increased temperature mainly due to the carrier mobility degradation. Although the gd degradation for L=28nm occurs for both caused by the channel length modulation, the p-type device presents a higher degradation in all range of temperature due to the gate to channel electrostatic coupling as already observed in the SS behavior (figure 3).

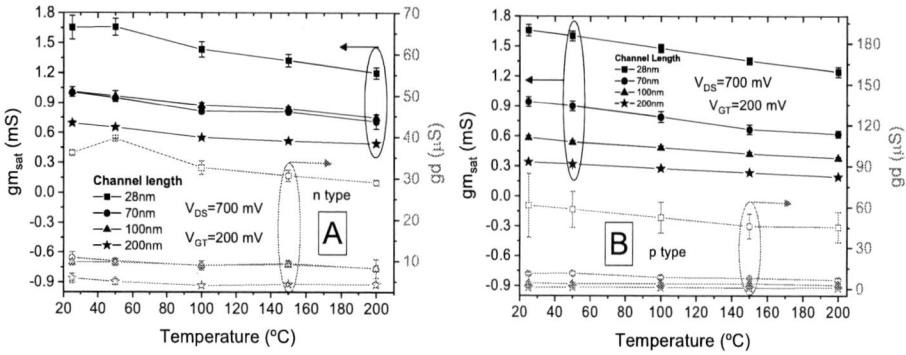

Figure 5. Transconductance (left) and output conductance (right) in saturation region for N-type (A) and P-type (B) GAA Nanosheet FETs as a function of temperature.

The intrinsic voltage gain (A_V) does not change significantly with temperature due to the compensation between the gm degradation and gd improvement as shown in figure 6. Long p-channel devices present higher A_V (gm/gd) than n-type GAA FETs due to the smaller values of gd which is not compensated by the gm degradation. However, for short devices the higher gd degradation results in smaller A_V.

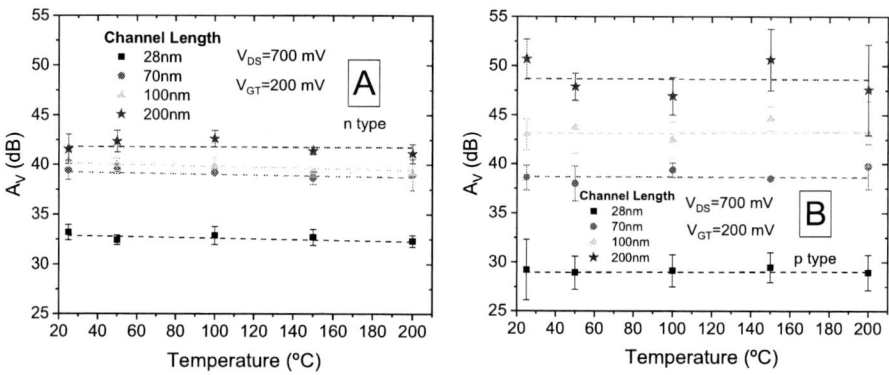

Figure 6. Intrinsic Voltage gain (A_V) in strong inversion for N-type (A) and P-type (B) GAA Nanosheet FETs as a function of temperature.

As said before, A_V is calculated considering gm_{sat} and gd and both depend on the mobility, so it is possible to note that A_V is barely affected by temperature. Other important highlight is the obtained A_V values for the studied GAA MOSFETs that are very high considering other multiple gate MOSFET structures. Both nMOS and pMOS nanosheet devices, with channel length of 100 nm, reach around 42 dB and 50dB respectively, while for bulk FinFETs , junctionless SOI nanowires and Ω-gate SOI nanowires the values achieved are 34 dB (6), 37 dB (7) and 40 dB (8), respectively, for the same V_{GT} and similar channel length.

Even though the longest devices show better SCE immunity and higher A_V values, the device with a channel length of 28 nm still shows a good performance. The GAA nanosheet achieves A_V of 33 dB and 28 dB, for n and p-type, respectively, while triple gate SOI nanowires, with channel length of 40 nm, and Ω-gate SOI nanowires, with channel length of 20 nm, presents values around 27 dB (9) and 20 dB (10), respectively.

Conclusion

In this work, the influence of temperature on basic parameters of the GAA nanosheet MOSFETs was analyzed. The V_{TH} showed low variation with temperature increase, reaching as low as -0.4 mV/°C. The SS values were very close to the theoretical limit and DIBL showed very low values, even when affected by high temperatures.

The devices achieved a maximum A_V of 50 dB and a minimum of 28 dB, for the pMOS longest L (200 nm) and shortest L (28 nm) respectively, which are higher than most of other MOS technologies around the same channel length. In addition, the devices were barely affected by temperature increase, since both gm_{sat} and gd are lowered by high temperatures.

Acknowledgments

The authors would like to thank CNPq and CAPES for the financial support.

References

1. A. Veloso, M. J. Cho, E. Simoen, G. Hellings, P. Matagne, N. Collaert, and A. Thean, ECS Trans., 72(2), 85 (2016).
2. D. Boudier, B. Cretu, E. Simoen, A. Veloso, and N. Collaert, Solid-State Electron., 143, 27 (2018).
3. J. P. Colinge, Silicon on Insulator: Materials to VLSI, Kluwer Academic Publishers, (1997).
4. S. Mukherjee, S. Dutta, J. Ghosh, D. Saha, A. Laha and S. Ganguly, 2019 Electron Devices Technology and Manufacturing Conference (EDTM), p. 118 (2019).
5. R. Keller, Total Ionizing Dose Effects in Silicon Bulk FinFETs at Cryogenic Temperatures (master), Faculty of the Graduate School of Vanderbilt University, Nashville, Tennessee (2017).
6. A. V. Oliveira, P. G. D. Agopian, J. A. Martino, E. Simoen and C. Claeys, 2014 International Caribbean Conference on Devices, Circuits and Systems (ICCDCS), p. 1 (2014).
7. R. Trevisoli, R. T. Doria, M. de Souza and M. A. Pavanello, EUROSOI-ULIS 2015, p. 265 (2015).
8. L. Almeida, P.G.D. Agopian, J.A. Martino, S. Barraud, M. Vinet and O. Faynot, 2016 IEEE SOI-3D-Subthreshold Microelectronics Technology Unified Conference (S3S), p. 1 (2016).
9. B. C. Paz, M. A. Pavanello, M. Casse, S. Barraud, G. Reimbold, M. Vinet and O. Faynot, EUROSOI-ULIS 2016, p. 170 (2016).
10. V. T. Itocazu, V. Sonnenberg, J. A. Martino, S. Barraud, M. Vinet and O. Faynot, EUROSOI-ULIS 2017, p. 192 (2017).

Electrical Behavior of Effects LCE and PAMDLE of the Ellipsoidal MOSFETs in a Huge Range of High Temperatures

E. H. S. Galembeck[a] and S. P. Gimenez[a]

[a] Department of Electrical Engineering, FEI University Center, São Bernardo do Campo, Brazil
egon@fei.edu.br

This paper presents the electrical behavior at high temperature-range of the effects present in the non-standard gate layout style for MOSFETs, in which they are capable to boost the electrical performance in relation to the standard rectangular MOSFET (RM) counterpart. These effects are named Longitudinal Corner Effect" (LCE) and "PArallel connection of MOSFETs with Different channel Lengths Effect (PAMDLE). In this study, we study the ellipsoidal layout style for MOSFET, taking into account the Bulk technology of CMOS CIs of 180nm from TSMC. This work is based on three-dimensional numerical simulations and we conclude that the LCE and PAMDLE effects are always active in all temperatures studied (300K-573K) and, consequently, they are capable of boosting the EM electrical performance remarkably in comparison to the RM counterpart (83% for saturation drain current and 86% for the maximum transconductance), regarding the same gate areas and bias conditions for a temperature of 573K.

Introduction

There is a growing interest, in the past few years, in new electronic systems that must operate at high temperature (T) mainly for automotive, space, aircraft, aerospace, petroleum industry electronics, etc. To reach this purpose new device architectures have been proposed, such as the Gate-All-Around (GAA) (1), Multigate SOI MOSFETs (2), etc., or the use of other semiconductor materials, such as Silicon Carbide (SiC) for MOSFETs (3). However, there is a new concept, still little explored by the integrated circuits (ICs) industries, regarding planar Bulk Complementary Metal-Oxide-Semiconductor (CMOS) ICs technologies, which is based on the simple change of the standard gate geometry of Metal-Oxide-Semiconductor Field Effect Transistor (MOSFET) a to non-standard (non-rectangular) one (4). This pioneering layout technique is capable of boosting the electrical performance of MOSFETs, thanks to two new effects of these structures, called "Longitudinal Corner Effect" (LCE) and "PArallel connection of MOSFETs with Different channel Lengths Effect" (PAMDLE) (4). One of the main attributes of this layout technique is that it does not add any extra cost to current planar CMOS ICs manufacturing process.

The first proposed innovative layout style has been the hexagonal gate geometry, named Diamond MOSFET (DM), which has been intentionally created to use the corner effect in the longitudinal direction of MOSFET's channel, i.e., it boosts the resultant longitudinal electric field (RLEF) and, consequently, its drain current (I_{DS}), transconductance, etc. Previous studies have reported remarkable gains in electrical performance of the Diamond layout style implemented in 1μm-Silicon-On-Insulator (SOI) in comparison to their standard Rectangular SOI MOSFETs (RSM) counterparts [same gate area (A_G), channel width (W) and bias conditions], from room to high temperatures (300K-573K), thanks to the LCE and PAMDLE effects, which are kept active in these operation conditions (5), (6).

Posteriorly, as an evolution of DM, the octagonal gate geometry implemented for MOSFET (Octagonal MOSFET, OM) has been created. The OM have been engineered in order to increase the breakdown voltage (BV_{DS}) and electrostatic discharge (EDS) tolerance (4). The OM, like DM, implemented in SOI technology (OSM) were experimentally studied in a huge range of temperature form 300K to 573K in relation to the RSM counterparts and the results indicated that OSM is capable of improving the saturation drain current (I_{DS_SAT}) with 159% and the unity voltage gain frequency (f_T) with 175%, for example (7)–(9). In order to eliminate the corners presents at octagonal gate geometry, it has been developed a MOSFET with an ellipsoidal gate geometry, named Ellipsoidal MOSFET (EM). This device boosts the BV_{DS} and the EDS robustness in relation to the OM counterpart, regarding the same gate areas (A_G), channel widths (W) and bias conditions (4).

Thus, in this scenario, this manuscript performs a comparative study by three-dimensional (3D) numerical simulations of the effects caused by the high temperatures conditions (temperature range from 300K to 573K) in the main the electrical parameters of the EM in relation to the RM counterpart, regarding the same A_G and bias conditions. Furthermore, for the first time, is analyzed the electrical behaviors of LCE and PAMDLE effects in the EM structure, regarding these conditions.

Device Characteristics

The EM is a natural evolution of OM and DM, respectively, in which it has been carefully engineered to eliminate the obtuse corners of the hexagonal and octagonal gate shapes. This simple change of gate geometry of MOSFET provides to improve the BV_{DS} and EDS robustness in relation to DM and OM. Fig. 1 presents the top schematic views of EM and RM counterpart with their respective geometric dimensions, regarding the same A_G and bias conditions, which illustrates the RLEF and its vector components in the channel regions at the point P. In Fig. 1, B and b are the largest and smallest channel lengths, respectively and L is the RM channel length.

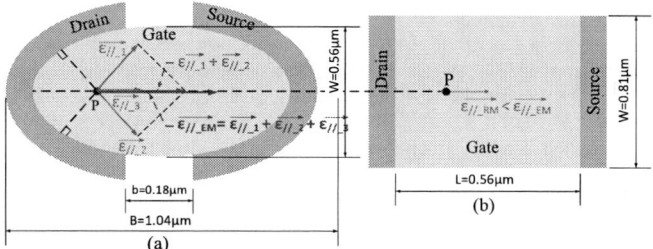

Figure 1. The top schematic views of EM (a) and its RM counterpart (b) illustrating their corresponding RLEF and their vector components.

The presence of LCE in the EM is capable of boosting the RLEF magnitude along its channel length ($\overrightarrow{\varepsilon_{//\_EM}}$) and it is always higher than observed in the RM counterpart (Fig.1b - $\overrightarrow{\varepsilon_{//\_RM}}$), because the $\overrightarrow{\varepsilon_{//\_EM}}$ is the vectorial sum of three longitudinal electric field (LEF) components ($\overrightarrow{\varepsilon_{//1}} + \overrightarrow{\varepsilon_{//2}} + \overrightarrow{\varepsilon_{//3}}$) considering a point P located on the line defined by the foci (F1 and F2) of the ellipse and out of this line, the $\overrightarrow{\varepsilon_{//\_EM}}$ is given by only the vectorial sum of two LEF ($\overrightarrow{\varepsilon_{//1}} + \overrightarrow{\varepsilon_{//2}}$), while the RLEF in RM counterpart is composed by only one LEF component ($\overrightarrow{\varepsilon_{//\_RM}}$) in all channel region (4).

Due to the LCE, the drift velocity of mobile charge carriers in the channel region of the EM increases and, consequently, its I_{DS} increases and it is always higher than that of the RM counterpart (4), regarding that they present the same A_G and bias conditions.

Besides, as the EM electrical circuit can be considered as the parallel association of infinite number of conventional MOSFETs with infinitesimal channel width and different Ls ($b \leq L \leq B$), the PAMDLE effect is responsible for reducing the effective channel length $\left\{L_{eff} = B/\left[\sin^{-1}\left(\sqrt{1 - b^2/B^2}\right)\right]\right\}$ of the EM in comparation to the L of an RM counterpart. This is justified because the EM I_{DS} tends to further flow in the edges of the channel region, because in the edges Ls are smaller than those in the center of the device. Thus, as the I_{DS} is inversely proportional to L, the EM I_{DS} tends to be higher than that of the RM counterpart (4). These two effects (LCE and PAMDLE) of the Ellipsoidal layout style act jointly and therefore they are capable of boosting the electrical performance of MOSFETs. It is important to highlight that the L of RM must be equal to $\pi B/4$ so that it presents the same A_G of the EM counterpart.

A first order analytical model for the EM I_{DS} (I_{DS_EM}) as a function of the I_{DS} of RM counterpart (I_{DS_RM}) is presented in equation [1] (4):

$$I_{DS\_EM} = G_{LCE} \cdot G_{PAMDLE} \cdot I_{DS\_RM} \qquad [1]$$

where G_{LCE} [$= \sqrt{2(1 + \cos \alpha)}$ for $0° < \alpha \leq 90°$ and $\sqrt{(2 + \cos \alpha)}$ for $90° \leq \alpha < 180°$] are the gains related to the LCE effect and G_{PAMDLE} $\left[= L/L_{eff} = \frac{\pi}{4}\sin^{-1}\left(\sqrt{1 - \frac{b^2}{B^2}}\right)\right]$ represents the gain provided by the PAMDLE effect (4).

Results and Discussions

The 3D numerical simulations were performed by using of the Sentaurus TCAD from Synopsys (10). The simulations of curves of the drain current (I_{DS}) as a function of the gate voltage (V_{GS}) and drain current (V_{DS}) of the MOSFETs studied in this paper were calibrated taking into account experimental data of MOSFETs manufactured with the 180nm CMOS manufacturing process of TSMC, regarding a temperature equal to 300K. The models used in the 3D numerical simulations were: High field saturation model that describes mobility degradation at high electric fields; Philips Unified Mobility model that describes the majority and minority carrier bulk mobilities; Enormal model that describes the mobility degradation at interfaces; Shockley–Read–Hall for the behavior of the generation-recombination process.

Table I illustrates the values of the threshold voltages (V_{TH}) as a function of T of EM and RM counterpart, regarding V_{DS} equal to 50mV.

TABLE I. V_{THs} for different temperatures of devices used in this study.

	T [K]						
	300	*323*	*373*	*423*	*473*	*523*	*573*
EM V_{TH} [V]	0.48	0.46	0.42	0.38	0.32	0.26	0.22
RM V_{TH} [V]	0.48	0.44	0.40	0.36	0.32	0.26	0.22

Based on Table I, we observed that the devices V_{THs} reduce as the T increases, because it has a dependence with the Fermi potential, which reduces at high temperature (11). Besides, the Table I shows that the V_{TH} of EM and RM are similar in all temperatures and, in first approximation, it reduces linearly by about 1mV/C° with the increasing T.

One of the MOSFET parameters which has been pointed out to be of fundamental importance for the stable circuit operation over a wide temperature range is the zero-temperature coefficient (ZTC) point. The ZTC point of a MOSFET is the bias of V_{GS} which ensures that I_{DS} remains constant even with the temperature variation. The ZTC conditions derive from the mutual cancelation of charge carriers mobility and V_{TH} dependencies with the temperature, i.e., there is a balance between the degradation of charge carrier mobility, in which it reduces I_{DS}, and the decrease in V_{TH}, which increases I_{DS}, both caused by the increase of the T, because I_{DS} is directly proportional

73

to charge carrier mobility and inversely proportional to V_{TH} (12). In this condition is defined a bias point called ZTC (V_{ZTC}; I_{ZTC}), wherein I_{DS} presents small temperature sensitivity (both effects can cancel each other at a certain bias point) (12).

Fig. 2.a shows the I_{DS} normalized by the aspect ratio, W/L, [I_{DS}/(W/L)] as a function of V_{GS} for different Ts of the EM and RM counterpart. Based on these results, we observe that the ZTC biases of the EM and RM (V_{ZTC}) are practically similar (difference of 4% between both), while the EM I_{DS}/(W/L) in the ZTC point is 114% higher than the one simulated for the RM counterpart, thanks to the LCE and PAMDLE effects present in the EM structure, which are responsible for boosting the I_{DS} according to equation [1]. Besides, Fig. 2.b illustrates I_{DS}/(W/L) as a function of V_{DS}, considering different temperatures. The EM I_{DS}/(W/L) in all temperatures studied, considering the triode and saturation regions, is always higher (about 1.8 times higher) than the one observed in the RM counterpart, due the LCE and PAMDLE effects.

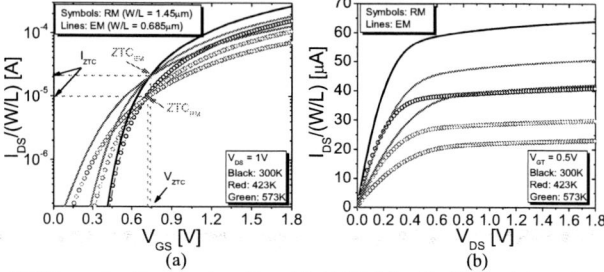

Figure 2. I_{DS}/(W/L) as a function of gate voltage (a), highlighting the ZTC points, and as a function of the V_{DS} (b), for different Ts for EM and the RM counterpart.

Based on these previous results, Fig. 3 presents the maximum transconductance normalized by the aspect ratio [gm_{max}/(W/L)] in Fig. 3.a and I_{DS_SAT}/(W/L) in Fig. 3.b as a function of the T.

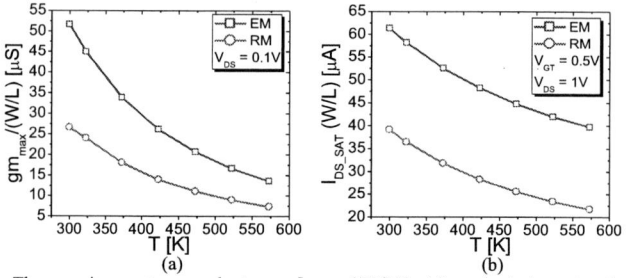

Figure 3. The maximum transconductance [gm_{max}/(W/L)] (a) and drain saturation current [I_{DS_SAT}/(W/L)] (b) as a function of temperature for both devices.

We have observed that the gm_{max}/(W/L) is always higher than the one found in RM counterpart for all temperatures studied, in which the gains are of 94%, 87% and 86% for 300K, 423K and 573K, respectively. Besides, according to the Fig. 3.b, the EM I_{DS_SAT}/(W/L) are also always higher than those found in the RM counterpart (it can be up to 1.8 times higher, depending on temperature), thanks to LCE and PAMDLE effects and they reduce with the temperature increases, because the electron mobility in the channel of MOSFETs decreases with the increase of the T (11). Therefore, based on these results, the EM can be regarded an alternate device to boost the electrical performance of the MOSFETs. Besides, taking into account the results shown above, we have concluded that the LCE and PAMDLE effects are always active in this huge range of high temperatures.

Fig. 4 presents the 3D numerical simulation results of the RLEF (color maps and lines) in the RM (Fig. 4.a) and EM (Fig. 4.b), respectively, regarding two different temperatures and for V_{DS} and V_{GS} equal to 1V. The RLEF in the center of channel and along the L, in average, are similar for both devices, as we can see in the Fig. 4.c (A-A' cut). The RM RLEF is constant along the W and in the center of channel, while the EM RLEF varies along the center of the channel and W. We also observed that the intensity of the RLEF in the edges of the gate region of the EM is higher (43% and 56% for T equal to 300K and 573K, respectively) than the one in the center of the channel [Fig. 4.c (B-B' cut)]. This is justified because the channel length in the edges of the EM are smaller than those in the center of the channel.

We can also see that the RLEF lines mapping is different among the devices. The RLEF lines in the RM are straight along the channel region (Fig. 4.a), while they are curved out of the center of the channel of the EM. This can be justified due to the LEF components along the channel length are perpendicular to the drain/source and channel region interfaces (DEPAMBBRE effect) (4).

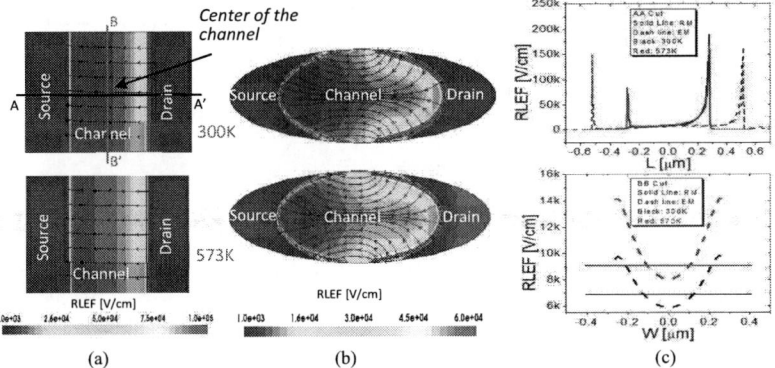

Figure 4. Horizontal cut-off in the RM (a) and EM (b) below 1nm of the channel/oxide interface, in saturation region ($V_{DS}=V_{GS}=1V$), highlighting RM and EM RLEF (colors map and lines). Besides, the RLEF intensities as a function L (A-A' cut) and W (B-B' cut) (c).

Fig. 5 illustrates the drain current density, J_{DS}, (color maps and lines) along the structure of the RM (Fig. 5.a) and EM (Fig. 5.b) obtained by 3D numerical simulations, regarding two different temperatures and considering the V_{DS} and V_{GS} equal to 1V. The drain current density in RM is homogenously spread along the W, as can be seen from the figure Fig. 5.a. However, the drain current density of the EM presents a maximum value (highlighted with red color in Fig. 5.b) in the edges, i.e., the drain current density increases from the center to the edges of the EM structure. Besides, the drain current density of the EM along L is higher than found in the RM counterpart (approximately 14% and 11% for 300K and 573K, respectively, regarding the center of the channel and along the L), as we can see in the Fig. 5.c (A-A' cut), except when the channel length of the EM becomes greater than the RM L. We have also observed that the drain current density in the edge of the gate region is always higher (approximately 51% and 73% for 300K and 573K, respectively) than that found along the entire channel region of the RM counterpart, according to Fig. 5.c (B-B' cut). Therefore, Fig. 5 proves the existence of the PAMDLE effect in EM structure.

The mapping of EM I_{DS} lines, illustrated by Fig. 5.b, shows us that EM J_{DS} lines are curved near the edges of the channel and straight (same RM behavior - Fig. 5.a) in the center of the L, due to the interactions of the longitudinal electric field components, i.e., EM I_{DS} lines follow the behavior of the RLEF lines, as illustrated in Fig. 4.b

The explanations related to Fig. 4 and Fig. 5 are in conformity with the theory presented in this paper and in reference (4), in which it proves that the LCE and PAMDLE effects are always active as the temperature increases and consequently these effects are capable of boosting the EM electrical performance in comparison to the one of the RM counterpart.

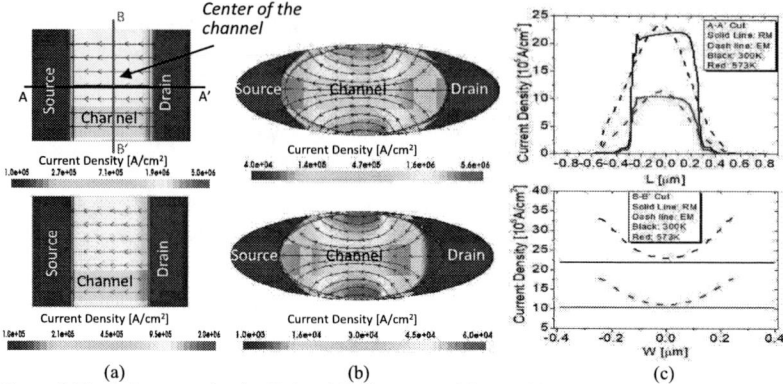

(a)   (b)   (c)

Figure 5. The drain current density (J_{DS}) and J_{DS} lines map of the RM (a) and EM (b) at 300K and 573K, regarding $V_{DS}=V_{GS}=1V$. Besides, the J_{DS} as a function of the L (A-A' cut) and W (B-B') (c).

Conclusion

In this work we analyzed the electrical behavior of the EM in relation to the RM counterpart (same A_G and bias conditions) at high temperature-range, based on 3D numerical simulations. Based on these results, we conclude that the EM always presents a better electrical performance {1.8 times higher for the saturation drain current normalized [$I_{DS\_SAT}/(W/L)$] and 94% higher for the transconductance normalized by the aspect ratio [$gm_{max}/(W/L)$]} than that found in the RM counterpart at high Ts. Besides, it was observed by 3D numerical simulations the behavior and existence of the LCE and PAMDLE effects in the EM structure. The results have shown that both effects are always actives as the temperature increases. Therefore, the ellipsoidal layout style can be considered as an alternative design technique in order to boost the electrical performance of MOSFETs at high-temperature conditions.

Acknowledgment

We thank to the grant #2017/10718-7, São Paulo Research Foundation (FAPESP) and to the grant #308702/2016-6, CNPq, for financial support.

References

1. M. Wahab, S. Shin and M. Alam, *IEEE Trans. Electron Devices*, **22** (11), 3595 (2015).
2. J.-P. Colinge, L. Floyd. A. Quin, G. Redmond, J. C. Alderman, W. Xiong, C. R. Cleavelin, T. Schulz, *IEEE Electron Device Lett.*, **22** (11), 172 (2006).
3. J. Qi, K. Tian, Z. Mao, S. Yang and W. Song, *IEEE APEC 2018*, 2712 (2018).
4. S. P. Gimenez, *Layout Techniques for MOSFETs*, Morgan & Claypool (2016).
5. E. Galembeck, S. P. Gimenez, *2013 IEEE S3S*, 978-1-4799-1361-9 (2013).
6. S. P. Gimenez, E. Gaembeck, *Microelectron. Reliab.*, **55** (2), 783 (2015).
7. S. P. Gimenez, E. Galembeck, *IEEE Trans. Device Mater. Reliab*, **15** (4), 626 (2015).
8. E. Galembeck, S. P. Gimenez, *IEEE Trans. Dev. Mater. Reliab*, **17** (1), 221 (2017).
9. E. Galembeck, S. P. Gimenez, *Integr. Circuits Syst.*, **14** (2), (2019).
10. Sentaurus Device User Guide (2018) Available: https://www.synopsys.com.
11. J.-P. Colinge and C. A. Colinge, *Physics of Semiconductor Devices*, (2002).
12. I. M. Filanovsky and A. Allam, *Circuits Syst. I Fundam. Theory Appl.*, (2001).

CHAPTER 2

Sensor and Biosensor

Post-CMOS Compatible Silicon MEMS Nano-Tactile Sensor for Touch Feeling Discrimination of Materials

H. Takao[1,2]

[1]Faculty of Engineering and Design, Kagawa University, JAPAN
[2]Nano-Micro Structure Device Integrated Research Center, Kagawa University, JAPAN
Correspondent author email: takao@eng.kagawa-u.ac.jp

Based on silicon post-CMOS compatible process, we have realized high performance silicon MEMS tactile sensor called "Nano-tactile sensor", which is comparable to the performance of fingertip sense of touch. In this tactile sensor device, all the mechanical structures are made from single crystalline silicon which is the active layer of SOI wafers. No elastomer/polymer structures are used in the mechanical sensing structure, and softness and elasticity necessary to tactile sensing are realized by silicon micro mechanical structures. In the latest version of our tactile sensors, six contactors with fingerprint-like shape are integrated at a pitch of 500μm to get high spatial resolution tactile images. Each fingerprint-like contactor reproduces vertical motion (by micro roughness) and horizontal motion (by frictional force) of a fingerprint closely under sweeping motion of fingertip in measurement. The performances are enough to analyze surface touch feelings of "Hair surface condition", "Skin condition", and "Touch feeling of various papers and clothes" like human fingertip. Combination of high-resolution tactile sensor and deep neural network is a strong approach to reproduce human fingertip sensation by state-of-the-art electron device technology.

Introduction

Semiconductor silicon has realized various kinds of valuable electron devices. CMOS technology is the present standard silicon process to fabricate integrated circuits and systems. Therefore, the word "Post-CMOS" has been an important word which means that the technology makes it possible to fabricate various kinds of MEMS sensors/actuators/passives with highly sophisticated modern CMOS devices. Also, it emphasizes that MEMS devices fabricated by post-CMOS technology has "high compatibility" with "silicon Fabs" that are the most universal and distributed fabrication facilities in the world. High performance devices and fine microstructures can be fabricated by universal post-CMOS fabrication processes at present. Post-CMOS compatible process is a key technology to realize ultra-high-performance silicon MEMS tactile sensors in this study. We, humans have very sophisticated sense of touch on our fingertip skin, and we can recognize and distinguish various and delicate difference of touch feelings obtained by "sweep motion" of fingertip on various kinds of materials and objects. Our fingertip skin has the highest density of force and vibration (mechanical) receptors like Meissner's corpuscles and Merkel disks under the surface skin layer where fine pitch patterns of fingerprint are formed on. It is known that human's fingertip has a very high spatial resolution below 100μm or less, and can recognize existence of 13nm-pitch patterns as reported recently. The high performance of our fingertip sense of touch has never been realized by any tactile sensor device. In order to reproduce artificial sense of touch like the fingertip, very high performances on "spatial resolution" and "input sensitivities" are required to integrate all on a small tactile sensor. Therefore, CMOS-compatible fabrication technology is the most suitable process to fabricate such high-performance tactile sensors with integrated functions.

Based on a silicon post-CMOS compatible process, we have realized high performance silicon MEMS tactile sensor called "Nano-tactile sensor", which is comparable to the performance of fingertip sense of touch. In this tactile sensor device, all the mechanical structures are made from single crystalline silicon which is the active layer of SOI wafers. No elastomer/polymer structures are used in the mechanical sensing structure, and softness and elasticity necessary to tactile sensing are realized by silicon micro mechanical structures. All the fine mechanical movements and sensing circuits with strain-sensitive diffusion resistors (i.e. piezoresistors) are designed by 2D-CAD, and the characteristics can be controlled by layout pattern sizes. The contactor parts of the tactile sensor have curved shape which is similarly designed to the cross-section of a fingerprint, and its suspension springs are designed to have similar spring constant with human's fingertip skin surface. In the latest version of our "Nano-tactile sensors", six contactors with fingerprint-like shape are integrated at a pitch of 500μm to get distributed tactile images at a high spatial resolution. The pitch of 500μm corresponds to the average value of human's fingerprint patterns. Each fingerprint-like contactor reproduces vertical motion (by micro roughness) and horizontal motion (by frictional force) of a fingerprint closely under sweeping motion of fingertip in touch feeling measurement. Spatial resolution of our tactile sensor reaches to sub-micron (200nm), and its force resolution of input reaches below 50μN range. These high performances are enough to reproduce part of our fingertip sense of touch. The Nano-tactile sensors can get touch feeling waveforms of "hair surface condition", "skin texture and the condition", and touch feelings of various kinds of "papers" and "clothes" at a high spatial resolution like our fingertip skin.

In the later sections, details of the very high-resolution tactile sensor are presented, and novelty and technologies are discussed.

Silicon MEMS Nano-Tactile Sensor based on post-CMOS Compatible Technology

A novel "array type" monolithic fingerprint-like tactile sensor has been developed with post-CMOS Compatible silicon MEMS technology, which is acquiring high resolution images of tactile information of surface. Six high resolution tactile sensors are arrayed in line at a pitch of 500μm, which is equivalent to typical pitch of human's fingerprint. Each tactile sensor can detect bi-axial motion of contactor for detection of "micro surface roughness" and "local slip friction", even if it is a very soft material like clothes. As an important result obtained with the array of tactile sensors, "non-linear deformation characteristic of soft fabrics" has been acquired successfully. Moreover, distributed tactile information on "shape" and "slip friction" has been 3D visualized on a soft cloth surface at a 100μm spatial resolution for the first time.

Artificial realization of human's fingertip sensation has been an important challenge for technologies such as human sensitivity evaluation and tactile presentation. In human fingertip, densely distributing tactile receptors can feel micro step less than 10μm and vibration of friction below 100μN. Even though fingertip skin can touch softly to softer materials, it can detect tactile information at multiple points, and its spatial resolution is high (below 100μm or less). A touch feeling evaluation device KES (Kawabata Evaluation System) has been used as standard evaluation method, and it measures mechanical characteristics of a sample surface in a square of a few centimeters. However, it cannot detect tactile information on micro region that human can feel, and its evaluation results do not always correspond to human touch feeling. Also, previous macro-scale tactile sensors employ piezoelectric polymer films such as PVDF [1] and pressure-sensitive rubber sheet, and they cannot divide the force into normal force and frictional force independently. Tactile sensors in MEMS field have realized multi-axis force detection [2-5], however, their sensitivity and spatial resolution (0.5~2mm) cannot compare with the ability of fingertip skin mentioned above. To obtain fine sensation like human fingertip, it is required to aim a sensor reproducing functions of fingertip skin sensation. High resolution two-axis MEMS tactile sensors with 100μm-spatial resolution and 50μN-sensitivity have been reported by our group [6, 7]. Although their sensitivity and spatial resolution are comparable to fingertip skin, they have only single contactor, and "multiple point detection" like fingertip skin surface has not been realized. So, a tactile sensor reproducing a structure of fingerprint has been developed in order to realize both high spatial resolution and multiple point detection like human's fingertip in this study. An array of highly sensitive six tactile sensors is monolithically integrated. Using the array function, "non-linear deformation characteristic of soft fabrics" has been acquired successfully, and distributed tactile information on "shape" and "slip friction" has been 3D visualized on a soft cloth surface at a 100μm spatial resolution.

Configuration of the Sensor

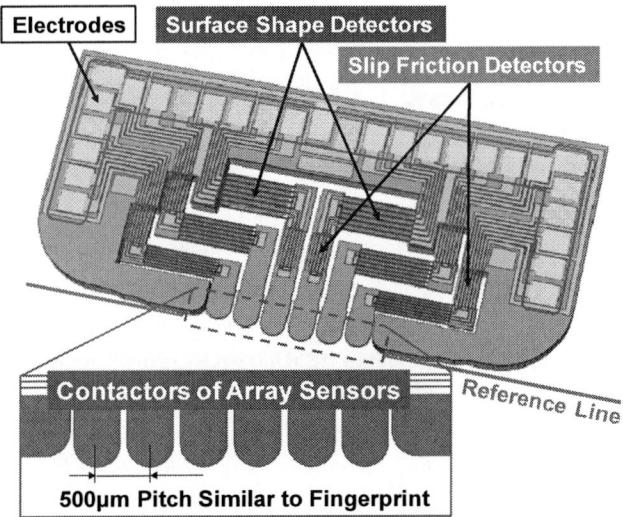

Figure 1: Configuration of monolithic fingerprint-like tactile sensor array.

Figure 1 shows the configuration of the fingerprint-like tactile sensor array integrating six independent bi-axial tactile sensors in line. It consists of the six contactors imitating human's fingerprint ridges, detection units of beams for independent detection of bi-axial motion, and chip frame working as the reference line for sweep motion. Devising the different shape of detection units, all the sensors are designed to have uniform sensitivity in the array. The six contactors are arrayed in 500μm pitch like pitch of typical fingerprint ridge, and the round shape of the tip is designed to realize similar cross-section of human's fingerprint. All the sensor structures are monolithically fabricated in a 50μm-thick SOI active layer, and the motion of contactors are detected by full-bridge piezoresistor circuits integrated on 16μm-width silicon beams. The circuits of full-bridge piezoresistors are fabricated by ion implantation process on the single crystalline silicon beams.

Operating motion of the integrated bi-axial tactile sensors on the sample object is illustrated in fig. 2. The device is pushed on the measuring surface first, then it is swept like human's touching motion. Sweeping the device on the sample surface, contactors move following the surface shape and slip friction generated at the contacting points. Vertical motion is detected by horizontal spring units as surface shape, and the horizontal motion is detected by vertical springs as slip friction.

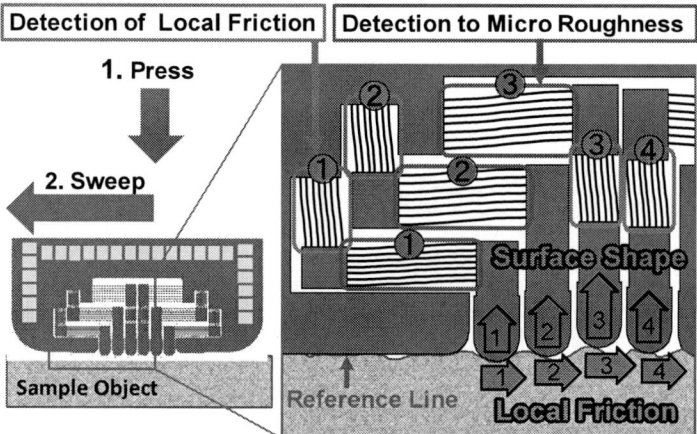

Figure 2: Operating motion of the integrated bi-axial tactile sensors on the sample object to be measured.

Post-CMOS Compatible MEMS Device Fabrication Process

In the device process, the circuit part is fabricated first, and temperatures of the following processes are below 400°C. Therefore, it is compatible with post-CMOS fabrication processes. Figure 3 shows entire view and cross-sectional view of the fabrication process flow of the tactile sensor. A p-type SOI wafer was used as the starting material. The device silicon layer is 50μm thick, the BOX (SiO2) layer is 0.5μm. First, the highly doped diffusion layer is formed by a thermal phosphorus diffusion process for circuit wiring, as shown in fig.3(a). Second, the piezoresistor parts are fabricated by ion implantation process of phosphorus and thermal annealing, as shown in fig 3(b). The width of piezoresistor is 4μm, and length is 120μm. Third, Cr films are sputtered on both the active layer surface and the backside surface of wafer. They are patterned to form hard mask for the following Deep-RIE. The first deep-RIE of the device layer is performed to fabricate the movable structures, as shown in fig.3(c). It is followed by the second deep-RIE process to etch the handle layer under the movable sensor structures, as shown in fig.3(d). Finally, the BOX layer is etched by HF solution to release the movable structures.

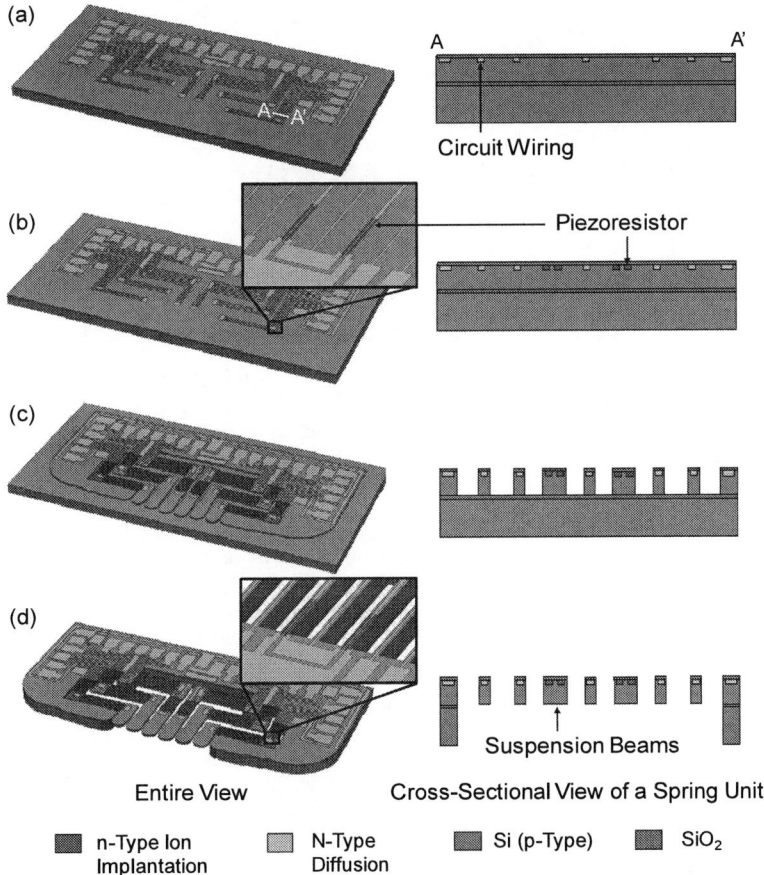

Figure 3: Post-CMOS Compatible Fabrication process of the Tactile Sensor Chip.

Figure 4 shows a photograph of the fabricated monolithic fingerprint-like tactile sensor array and the details of the chip. As shown in the photographs, the contactor array, the 16μm-width silicon suspension beam units integrating piezoresistors are successfully fabricated as designed.

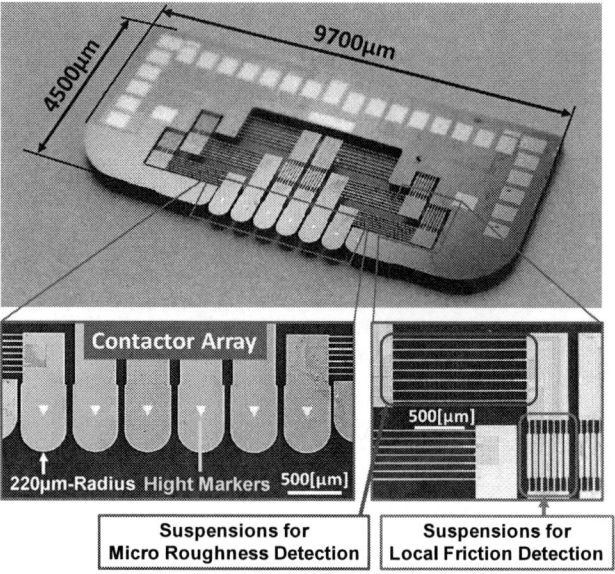

Figure 4: Chip photograph and details of the monolithic fingerprint-like tactile sensor.

Performances of the Nano-Tactile Sensors

In this section, performances of the nano-tactile sensor device is discussed. The sensor packaging and electrical setup were as follows. The sensor chip was bonded and fixed on a pitch converter board. Then electrical connection between the electrodes of the sensor and the board was performed by bonding Al wire. Integrated Wheatstone bridge circuits were connected to a data logger through an instrumentation amplifier circuit. The supply voltage of the device was 4.0V, and the gain factor of the amplifier circuit was 100. Responses of the tactile sensors to bi-axial inputs were precisely evaluated using a micromechanical testing and assembly system (FemtoTools FT-MTA03) with resolution of nN range. Since this testing device can simultaneously measure displacement and force, it is possible to measure the sensor sensitivity and the spring constant of the detection units. The response of the sensor to vertical displacement (height of surface shape) was measured pressing a probe of the testing device in vertical direction. Response to horizontal force (slip friction) was measured pressing the probe in horizontal direction.

Measured relationships between vertical displacement and two axis output signals are shown in fig. 5(a). Measured relationships between horizontal input force and two axis output signals are shown in fig. 5(b). The sensor output signals proportionally respond to the displacement and force with high linearity. The cross-sensitivities are low as below 3 % thanks to the structure of independent detection units. Resolutions of the sensors were 0.2μm in vertical displacement and 10μN in horizontal slip friction, respectively. Measured spring constants were approximately equal to the designed values, and their variation was within 7% in the integrated six contactors.

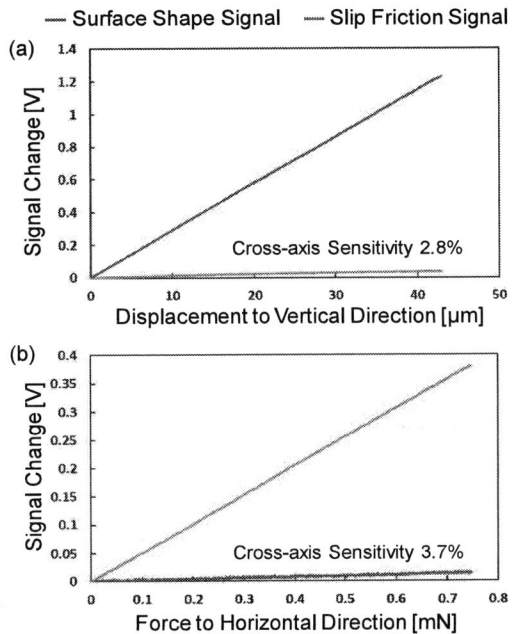

Figure 5: Response of tactile sensor to vertical displacement (a) and horizontal load (b).

Fig.6 (a) shows the experimental setup of device evaluation. The sensor device is fixed to the opposite side of the measured sample. The measured sample is fixed on the one-axis linear motion stage controlled by a measurement PC. The contactor tip and the reference plane were pushed to the sample surface, and the sample on the movable stage was swept at a constant speed. The sweeping speed was controlled to be 1mm/sec in this experiment.

Fig.6 (b) shows a video frame shot in the sweeping motion on "Plain Stitch" cloth sample. During the sweeping motion, the contactors moved up and down independently according to the surface shape of soft plain stitch. At the same time, sideways motion of the contactors generated by local slip friction at the contactor tip was observed. The contactors did not interfere each other, since the spring constant of the horizontal spring units was properly configured.

Fig.6 (c) shows all the signals obtained from the six tactile sensors on the measured cloth sample. All the signals are almost the same at the same position on the surface, and the weaving structure and slip friction on soft fabric were clearly detected. The six waveforms are well corresponding since all the tactile sensors are designed to have the same sensitivity. Each device realizes 100μm-spatial resolution and 50μN-sensitivity that are comparable to human's fingertip skin.

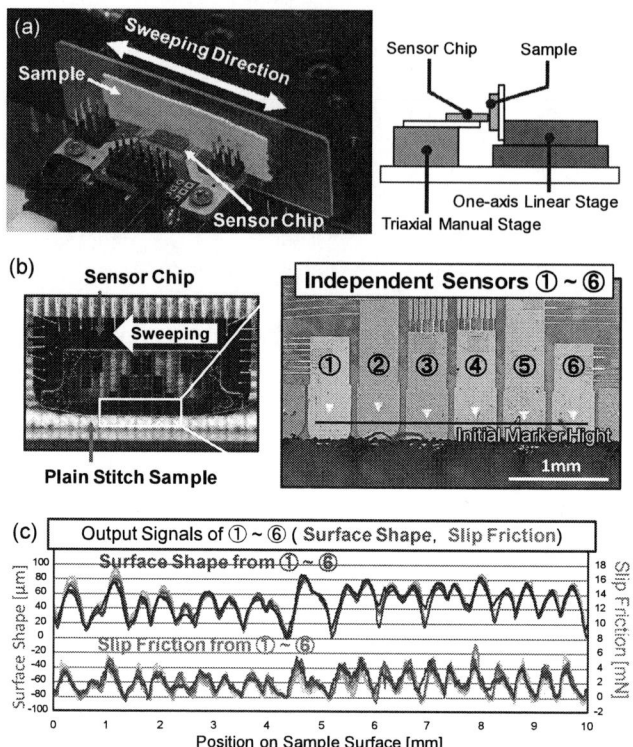

Figure 6: (a) Experimental setup of device evaluation. (b) A video frame shot in the sweeping motion on "Plain stitch" cloth. (c) All the signals from the six tactile sensors.

Fig.7 (a) shows result and experimental setup for measurement of "non-linear deformation characteristic" in soft fabric. In this experiment, sweeping the contactor array with a fixed pitch angle, pressing force to the sample varies gradually by six steps in the array. Since the object is deformed according to the pressing force, softness and non-linearity in deformation of soft material can be measured by the stepped position of the tactile sensor array in this study.

Fig.7 (b) shows a measurement result with a pitch angle of 2° as an example of evaluation. Comparing all the waveforms, non-linear increase of deformation and friction to the stepped squeeze depth is clearly characterized. Right side sensor in the array has a deeper squeeze depth than the left side sensor. So, signals of surface shape and slip friction appear stronger in the right side than the left side sensor in sweeping motion. Squeeze depth on the sample surface can be calculated from difference between each surface signals. It is considered that the difference in squeeze depth depends on softness of the measured sample. This result can be used for modeling of non-linearity in tactile sense of compressive elastomers and fabrics.

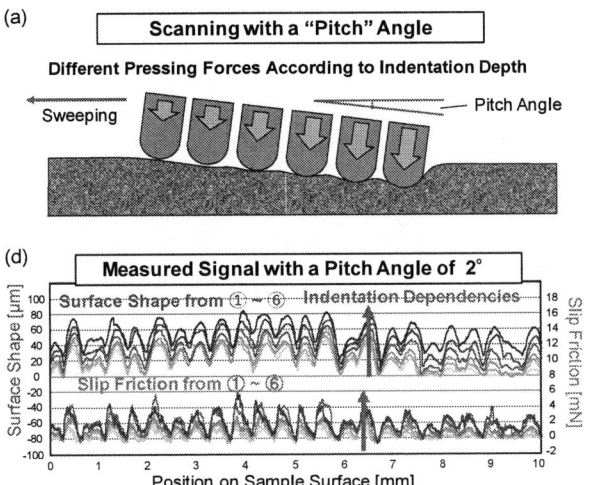

Figure 7: Result and experimental setup for measurement of "non-linear deformation characteristic" in soft fabric. Pressing force to sample varies gradually in the array.

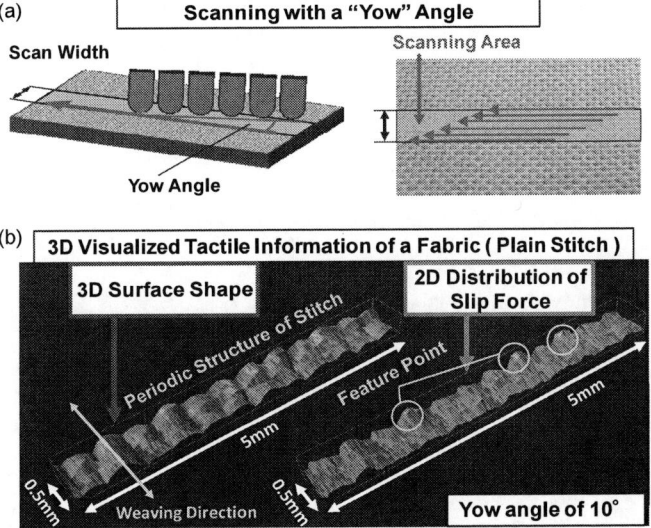

Figure 8: Distributions of tactile information on a 2D fabric surface plane. Stitch and weaving are visualized at a spatial resolution of 100μm. In addition, local friction corresponding to the surface shape is correlated successfully.

Fig.8 shows measured distributions of tactile information on a 2D fabric surface plane. The measurement was performed by sweeping the contactor array at a yow angle of 10° as shown fig.8 (a). The scanning part becomes a linear line, if scanning motion is performed without a yow angle (0°). Making a yow angle, it is possible to measure distribution of tactile information on six lines with a scanning width. The yow angle defines the scanning width and its spatial resolution (i.e. pitch) in the direction of the width, since the pitch of scanning lines depends on the yow angle. The minimum pitch of scan is 50µm, which is the thickness of the contactor. In this experiment with a yow angle of 10°, stitch and weaving are visualized at a spatial resolution of 100µm as shown in fig.8 (b). Periodic structure of the stitch and the directionality of weaving texture can be observed clearly. In addition, the characteristic local friction pattern corresponding to the surface shape is correlated each other successfully. There are some feature points where friction is stronger in the distribution of frictional force, and it is considered these are points feeling sticking on the fabric in human touching. Both results explained in Figs. 7 and 8 can be precisely obtained only by the monolithically integrated high resolution sensors.

Application of Deep Neural Network to Nano-Tactile Signals

Machine learning based on deep neural network (DNN) has become very important for sensing data analysis. We think DNN is very important to understand/reproduce human sense of touch since a trained DNN behaves similarly with human's senses based on our experiences. Since a lot of sample data are required for such applications, we have developed a measurement system called "Touch-Feeling Scanner". Fig. 9 shows the developed tactile scanner device integrating a nano-tactile sensor with single contactor. A number of tactile sensing data can be measured by the touch-feeling scanner integrating a nano-tactile sensor device. Obtained signals with the scanner were applied to train a DNN for discrimination of different kinds of clothes. 10 kinds of cloth samples have been successfully discriminated at a correct percentage over 95% as an example. Combination of high-resolution Nano-tactile sensors and state-of-the-art machine learning (DNN) is a strong approach to reproduce human fingertip sensation for various valuable applications.

Figure 9: Developed "Touch Feeling Scanner" integrating a nano-tactile sensor.

Conclusions

High performance silicon MEMS tactile sensor called "Nano-tactile sensor" has been realized based on post-CMOS fabrication process of MEMS, which is comparable to the performance of fingertip sense of touch. Six high resolution tactile sensors are arrayed in line at a pitch of 500μm, which is equivalent to typical pitch of human's fingerprint. Each tactile sensor detects bi-axial motion of contactor caused by "micro surface roughness" and "local slip friction". The signal responses of the fabricated sensor device showed good linearity. Also, the resolutions of surface shape height and slip friction were 0.2μm and 10μN, respectively. Non-linear deformation characteristic of soft fabrics has been acquired successfully. Moreover, distributed tactile information of "shape" and "slip friction" has been 3D visualized on a soft cloth surface at a 100μm spatial resolution. Combination of high-resolution tactile sensor and deep neural network is a strong approach to reproduce human fingertip sensation by state-of-the-art electron device technology.

Acknowledgments

This research has been supported by JST-CREST research funding program (Grant Number JPMJCR1531), and JSPS Grant-in-Aid for Scientific Research (A) (Grant Number 17H01271).

References

1. T. Okuyama, M. Hariu, T. Kawasoe, M. Kakizawa, H. Shimozu, and M. Tanaka, "Development of tactile sensor for measuring hair touch feeling," Microsystem technologies, vol.17(5-7) pp.1153-1160 (2011).
2. H. Takao, M. Yawata, K. Sawada, M. Ishida, "A multifunctional integrated silicon tactile imager with arrays of strain and temperature sensors on single crystal silicon diaphragm," Sensors and Actuators A Physical, vol. 160, pp. 69-77 (2010).
3. M. Sohgawa, D. Hirashima, Y. Moriguchi, T. Uematsu, W. Mito, T. Kanashima, M. Okuyama, H. Noma, "Tactile sensor array using microcantilever with nickel–chromium alloy thin film of low temperature coefficient of resistance and its application to slippage detection," Sensors and Actuators A: Physical, Vol. 186, pp. 32-37 (2012).
4. N. Thanh-Vinh, N. Binh-Khiem, H. Takahashi, K. Matsumoto, I. Shimoyama, "High-sensitivity triaxial tactile sensor with elastic microstructures pressing on piezoresistive cantilevers," Sensors and Actuators, A Physical, vol. 215, pp. 167-175 (2014).
5. Y. Hata, Y. Nonomura, Y. Omura, T. Nakayama, M. Fujiyoshi, H. Funabashi, T.Akashi, M.Muroyama, S. Tanaka, "Quad-seesaw-electrode type 3-axis tactile sensor with low nonlinearities and low cross-axis sensitivities," Sensors and Actuators A: Physical, vol. 266pp. 24-35 (2017).
6. K. Watatani, R. Kozai, K. Terao, F. Shimokawa, and H. Takao, "A "micro-macro" integrated planar MEMS tactile sensor for precise modeling and measurement of fingertip sensation," IEEE MEMS 2017, Las Vegas, USA, pp. 223-226 (2017).

7. N. Nakashima, K. Watatani, K. Terao, T. Suzuki, F. Shimokawa, H. Takao, "Sence of touch in submicron region realized by two-axis tactile sensor with a needle-like contactor," IEEE MEMS 2018, Belfast, UK, pp.870-873 (2018).

CMOS-MEMS Microgravity Sensors and Their Application

K. Masu, K. Machida, D. Yamane, H. Ito, N. Ishihara,
T. M. Chang, M. Sone, R. Shigeyama, T. Ogata, and Y. Miyake

Tokyo Institute of Technology
4259 Nagatsuta, Midori-ku, Yokohama, Kanagawa 226-8503, Japan

This paper reviews the feature of newly developed CMOS-MEMS
(microelectromechanical systems) and MEMS accelerometers and
its application for a diagnosis of Parkinson's disease. In order to
realize micro-G (1G=9.8m/s^2) level sensing, we propose and
develop capacitive MEMS accelerometers with Au proof mass using
the multi-layer metal technology. In order to reduce Brownian noise
(B_N), which determines the sensitivity of MEMS accelerometer, we
utilize Au as a high density proof mass. In addition, we investigate
the crystal structure of Au in terms of the device reliability.
Furthermore, we demonstrate the validity of applying the MEMS
accelerometer for the early-stage diagnosis of Parkinson's disease.
The experimental results regarding the sensor, the material and the
diagnosis suggest that our microgravity sensor can pave the new way
for the early-stage diagnosis of Parkinson's disease.

Introduction

MEMS (microelectromechanical systems) accelerometers are now widely used for a
variety of applications in automotive, industrial, healthcare, entertainment, consumer
handheld electronics, and so on (1, 2). The technical progress such as IoT (Internet of
Thing) and AI demands the micro-G (1G=9.8m/s^2) level sensing with small sensor size (3,
4). In order to realize the micro-G level sensing, MEMS accelerometers with high
sensitivity have to be developed. The sensitivity of the MEMS accelerometer depends on
the Brownian noise (B_N), which is inversely proportional to the mass of the proof mass (5-
8). In conventional silicon MEMS accelerometers, there are many reported techniques
using bulk-micromachining and/or wafer bonding (9-17). MEMS accelerometers
developed by bulk micromachined technology have B_N below 1 μG/√Hz and become quite
large sensor module (6, 7, 13, 14). B_N of MEMS accelerometer fabricated by surface
micromachined technology cannot exhibit the lower value than 10μG/√Hz (18-20). In order
to realize the high sensitivity, we have proposed capacitive MEMS accelerometers with
high-density proof mass and have been developing CMOS-MEMS and MEMS
accelerometers (21-24) using a post-CMOS process compatible to CMOS LSI process (25).

From the viewpoint of the usage of gold material for MEMS structures, we have
analyzed Au crystal structure to ensure the device reliability. It is revealed that the
electroplated Au material has a potential of the material strength. Furthermore, we have
developed diagnosis of Parkinson's disease as the application of the high sensitivity MEMS
accelerometer. So far, walking trajectory and postural abnormality of Parkinson's disease
patients and healthy persons are analyzed based on machine learning technique.

In this paper, we first present our recent research progress of high sensitivity MEMS
and CMOS-MEMS accelerometers. Next, we show the results of the crystal structure

analysis to the electroplated Au. Lastly, an application of the MEMS accelerometer is discussed in terms of early-stage diagnosis of Parkinson's disease.

MEMS Accelerometer for Micro-G Level Lensing

This chapter describes the design concept and fabrication results of the proposed MEMS accelerometer for micro-G level sensing.

Design concept

Figure 1 shows a schematic view of typical CMOS-MEMS and MEMS accelerometers (21-24). The MEMS accelerometer consists of suspensions, stoppers, a proof mass as upper electrode, and a fixed electrode. When the upper electrode is moving to the vertical direction by the input of acceleration, the capacitance is detected by the fixed electrode. The stoppers are built to prevent mechanical destruction caused by over-swinging of the upper electrode. The Brownian noise B_N is given by

$$B_N = \frac{\sqrt{4k_B T b}}{m} \ , \tag{1}$$

where k_B, T, b and m are the Boltzmann constant (1.38×10^{-23} J/K), the absolute temperature, the viscous damping coefficient, and the proof mass of an accelerometer, respectively (21). As shown in eq. [1], B_N is inversely proportional to the proof mass. Thus, the conventional Si MEMS capacitive accelerometers result in a large-sized Si proof-mass when a low B_N is

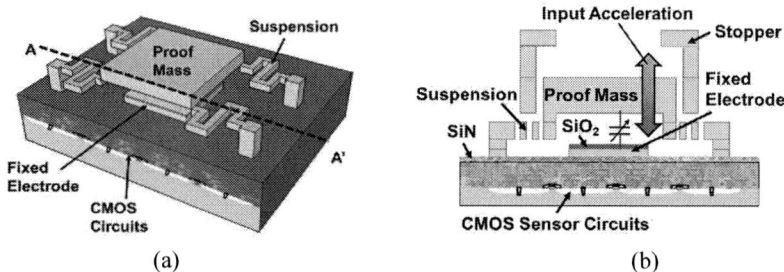

(a)   (b)

Figure 1. A schematic view of CMOS-MEMS accelerometer(a) and (b).

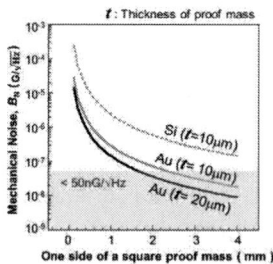

Figure 2. Mechanical noise analysis results.

obtained. Here, the density of Au (19.3×10^3 kg/m³ at 298K) (26) is nearly an order of magnitude higher than that of Si (2.33×10^3 kg/m³ at298K) (27). Utilizing gold material as a proof mass can achieve lower B_N compared to Si proof mass of the same size. For micro-G level sensing, the target resolution was determined to be below 50nG√Hz. Figure 2 shows analysis results of B_N as a function of the proof-mass size, where T and b are set to be 298K and 1.85×10^{-5} N·s/m, respectively. The thickness of the Si proof mass is set to be 10μm as the maximum thickness used for fabricating surface-micromachined MEMS accelerometers (27-29). We have developed gold proof mass structures with the thickness upwards of 12μm (22). On the other hand, we have to consider the structure of the proof mass comprising of the etching hole for removing the sacrificial film. Thus, the proof mass thickness will be designed by that of over 10μm. As the thickness of a single gold layer in the multi-layer technology was thinner than 20μm, we proposed to utilize multiple gold layers for the proof mass. The proof-mass structure with the thickness of 20μm can be achieved by using two gold layers in the multi-layer metal technology.

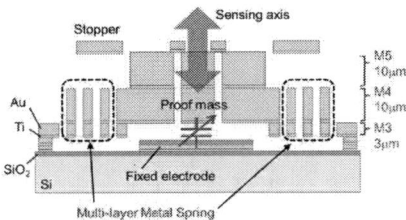

Figure 3. Schematic image of the proposed single-axis MEMS capacitive accelerometer.

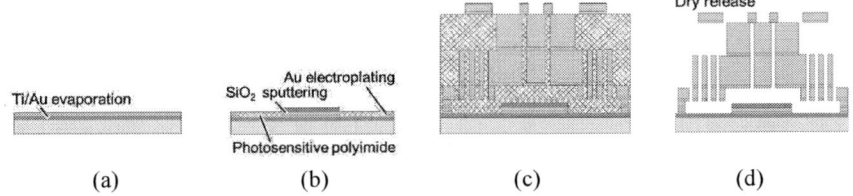

Figure 4. Process flow. (a) Seed layer deposition, (b) M1 and SiO₂ layers patterning, (c) M2-M6 patterning, and (d) sacrificial layer etching.

Device fabrication

Figure 3 shows a schematic image of the proposed single-axis MEMS capacitive accelerometer. We utilize the third (M3) and the fourth (M4) layers for the spring structure, and M4 and the fifth (M5) layers for the proof mass. Figure 4 describes the device fabrication process flow. Firstly, Ti/Au seed layers were deposited by evaporation on a thermal-SiO₂ as shown in Fig. 4 (a). Gold electroplating process was then used to increase the thickness of the first Gold layer (M1). Photosensitive polyimide as a sacrificial layer was spin-coated and annealed at the temperature of 310 °C. An SiO₂ layer with a thickness of 1μm was deposited by sputtering (Fig. 4(b)). With the same Au pattering as for M1 layer, we made another five Au layers (M2-M6), as illustrated in Fig. 4(c). Finally, all the sacrificial layers were removed by O_2 plasma etching (Fig. 4 (d)).

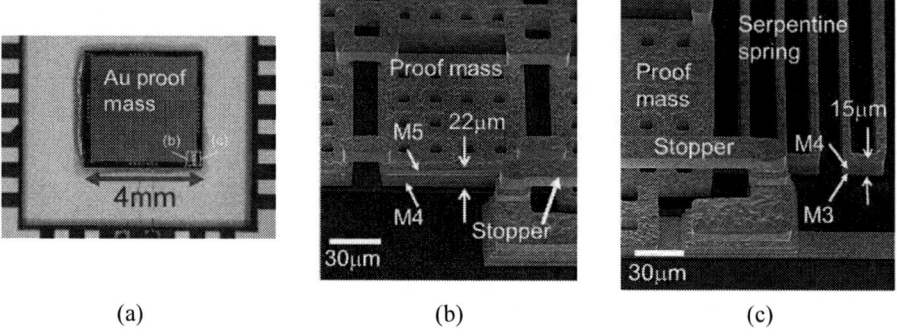

(a)   (b)   (c)

Figure 5.   Fabricated device. (a) Chip view, (b) proof mass and (c) spring structure.

Figure 5 (a) is a chip view of the developed MEMS accelerometer. A gold proof mass was fabricated on a silicon substrate with the footprint of 4mm×4mm. The SEM micrograph of the proof mass is shown in Fig. 5(b). The gold proof-mass structure with the thickness of 22μm was successfully developed by employing M4 and M5 layers. Figure 5 (c) shows the SEM image of the serpentine spring structure made of M3 and M4 layers. Stopper structures were made of M6 layer and set above the proof mass. The serpentine springs and the stoppers were placed at each corner of the proof mass.

Experimental results and discussion

Acceleration responses.   We measured capacitance between the proof mass and the fixed electrode when Z-axis acceleration was applied to the device as shown in Fig. 6.

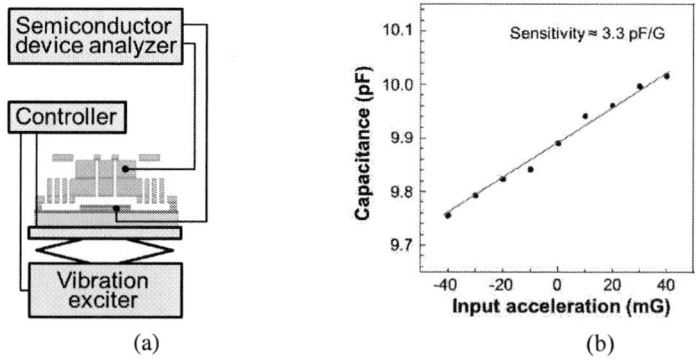

(a)   (b)

Figure 6. Measured capacitance as a function of input acceleration. (a) Experimental setup and (b) measurement results.

Figure 6 (a) is the experimental setup, where the packaged MEMS device was set on a custom-designed printed circuit board (PCB) on a vibration exciter (WaveMaker05, Asahi Seisakusho) with the minimum input-acceleration step of 0.01G. A semiconductor device analyzer (B1500A, Agilent Tech., Inc.) was used to supply the DC bias voltage of 0.5V and measure the capacitance change with a ±0.1-V sinusoidal voltage at the frequency of 300kHz. Figure 6(b) shows the measured capacitance as a function of input acceleration at the frequency of 49.9Hz. The sensitivity was experimentally obtained to be 3.3pF/G.

<u>Frequency characteristics.</u>   For the evaluation of mechanical characteristics and B_N, we measured the frequency responses of the fabricated MEMS device. Figure 7(a) shows the experimental setup with an LCR meter (IM3533-01, HIOKI E.E. Corp.). As shown in Fig. 7(b), the resonant frequency of the device was found to be 202Hz with the DC bias voltage of 0.5 V. To evaluate the B_N of the device, the following relationship (21) was used:

$$B_N = \sqrt{\frac{4k_B T \omega_r}{mQ}} ,$$

[2]

where ω_r and Q are the resonant angular frequency and the quality factor, respectively. Table 1 summarizes the design and measured characteristics of the device, and the actual B_N was obtained to be 22nG/√Hz. Figure 8 compares B_N against the capacitance sensitivity with conventional MEMS accelerometers. Owing to the high density of gold, the B_N of this work was more than an order of magnitude lower than those of conventional when compared with the same sensitivity performance.

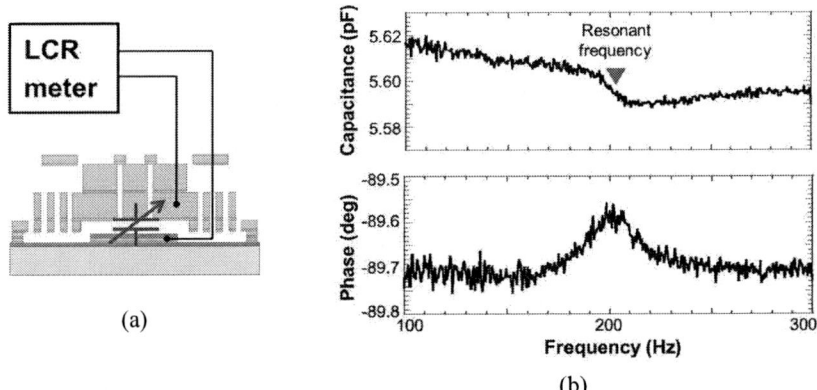

Figure 7. Measured capacitance and phase as a function of frequency. (a) Experimental setup and (b) measurement results.

Table 1 Summary of device characteristics.

	Design	Measured	Unit
m	3.28×10^{-6}	3.62×10^{-6}	kg
f_{res}	260	202	Hz
Q	289	131	
B_N	17	22	nG/√Hz

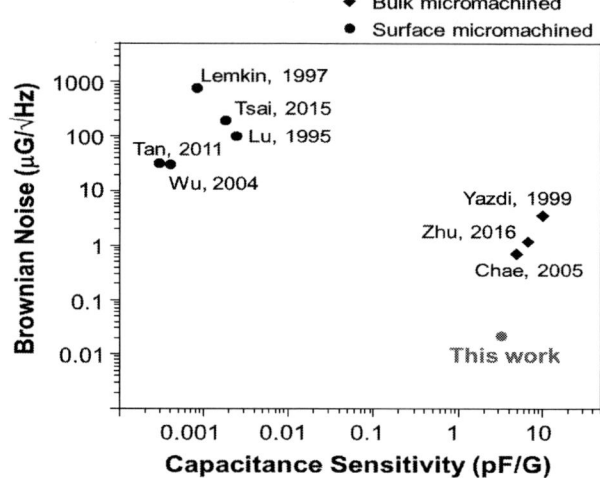

Figure 8.   Comparison of B_N versus capacitance sensitivity (8,9,17-19,30-32).

The material strength analysis of the electroplated Au

This chapter describes the material strength analysis based on the micro-mechanical property anisotropy of the electroplated Au film.

Experimental

The Au film was electroplated using a cyanide-based electrolyte (33). The Au film was electroplated on a Pt substrate. For the micro-compression test, two Au micro-pillars were fabricated by focused ion beam (FIB, Hitachi: FB2100). Details of the electroplating and the FIB fabrication are reported in a previous study (33). Dimensions of the Au pillar were $10 \times 10 \times 20 \mu m^3$ with a square cross-section, and the long-side was either perpendicular (pillar 1, Fig. 9(a)) or parallel (pillar 2, Fig. 9(b)) to growth direction of the Au film.

The compression test was conducted using a test machine specially designed for micro-sized specimens equipped with a flat-ended diamond indenter at a constant displacement of 0.1μm/s controlled by a piezo-electric actuator. The load resolution was 10μN.

Crystal structure of the Au film was evaluated by an X-ray diffractometer (XRD, Ultima IV, Rigaku), a scanning electron microscope (SEM, S-4300SE, Hitachi) equipped with electron back scatter diffraction (EBSD) function, and transmission electron microscopy (TEM, 200kV, JEM2100, JEOL). Deformation behavior of the specimens was observed by a scanning ion microscope (SIM) equipped in the FIB.

Results and Discussion

Microstructures of the Au micro-pillars before and after the micro-compression test are shown in Figs. 9(c) to (f). Columnar texture-like microstructures having the long-axis perpendicular to the long-side of pillar 1 were observed. The long-axis of the micro-columnar textures were parallel to the long-side of pillar 2.

After the compression test, brittle fracture along the texture boundary was observed in

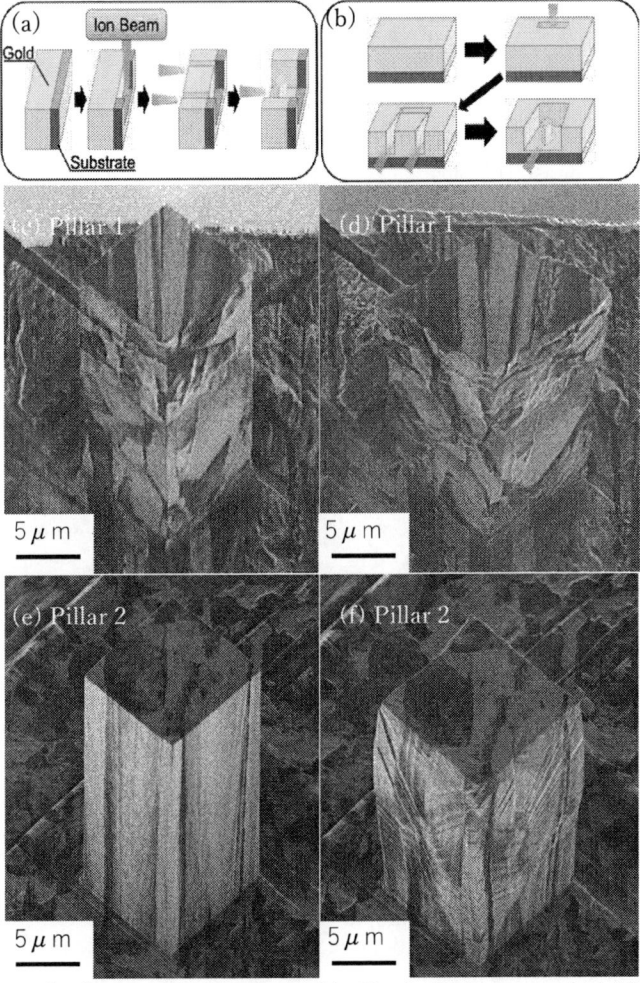

Figure 9. Schematic views of (a) pillar 1 and (b) pillar 2. SIM images of (c) (d) pillar 1 and (e) (f) pillar 2 before and after compression test.

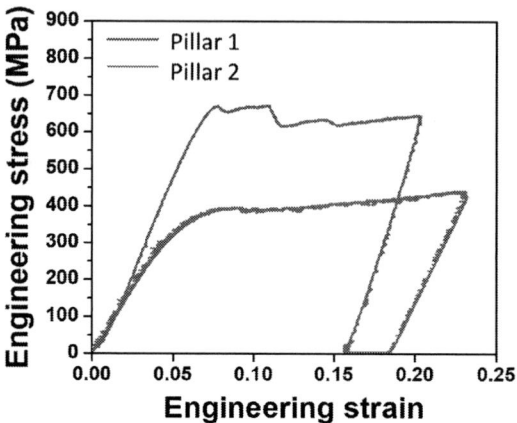

Figure 10. Engineering stress-strain curves.

pillar 1. In contrast, deformation with slip lines was observed in pillar 2, and it was generally defined as multiple slip deformation (34). Figure 10 shows the engineering stress-strain curves. Load drops and non-continuous change of the slope were observed in the plastic region for pillar 1, while the slope remained almost constant in the plastic region for pillar 2. Considering the deformation behavior of pillar 1, the load drop and the change in the slope are attributed to the brittle fracture. Pillar 1 showed a higher σ_Y (ca. 650MPa) than that of pillar 2 (ca. 300MPa). These anisotropies in the deformation behavior and the σ_Y can be explained by the information obtained from the microstructure analysis. The XRD pattern showed typical result of FCC Au and strong intensity for the peak of [200] orientation, shown in Fig. 11(a). Grain size of the Au film was estimated using the XRD result and the Scherrer equation. The grain size was estimated to be 14.7nm. From EBSD mapping with respect to the transverse direction (TD, Fig. 11(b)), the normal direction (ND, Fig. 11(c)), and the reference direction (RD, Fig. 11(d)), the Au film was confirmed to be composed of columnar grains, where the long-axis is parallel to growth direction of the Au film, and all grains showed [001] direction. This orientation is consistent with the XRD result, which the dominant peak is the peak of [200] orientation. On the hand, the grain size calculated from the EBSD mapping images was about 1.34μm, which is much larger than the value obtained from the Scherrer equation. The difference observed in the grain size was further clarified through TEM observation. As shown in Fig. 11(e), columnar grains were observed. Width of the grains is ca. 16~40nm, which corresponds well with the grain size calculated using the Scherrer equation. From the XRD, the EBSD mapping, and the TEM results, we can summarize that the Au film is composed of columnar textures with an average width of 1.34μm, and each texture is composed of columnar grains having an average width at about several tens of nanometer. In addition, the micro-columnar textures and the nano-columnar grains have the same crystal orientation, and the grain-like microstructures observed in EBSD mapping are confirmed to be textures. The anisotropy observed in the deformation behavior and the σ_Y can be explained from the columnar microstructures. For pillar 2, multiple slip deformation often occurs when FCC single crystal was compressed along [100] direction (34). Grains of the Au film are oriented in [100] direction. The grains in pillar 2 were compressed along the long-axis, which is the [100] direction. This led to formation of the slip lines shown in Fig. 9(f). Regarding the difference in the σ_Y, for pillar 1, the compression direction is perpendicular to the long-axis of the nanograins, which are orientated in [100] direction. Hence the compression can be treated as compression of grains with random orientation. In contrast, pillar 2 was compressed along [100] direction. In this compression direction, the compression can be deemed as compression of assemblies of columnar single crystals having the same crystal orientation. Because of the difference in the compression conditions, the relationship between critical resolved shear stress τ_{CRSS} and σ_Y is different for pillar 1 and 2. For compression of randomly oriented crystals (pillar 1), the relationship is represented by the Taylor model (35):

$$\sigma_{Y,Pillar\ 1} = M\tau_{CRSS} \ , \tag{3}$$

where, M is the Taylor factor, and the value is about 3.1 for FCC metal. For compression of assemblies of single crystals (pillar 2), the relationship is represented by the Schmid model [36]:

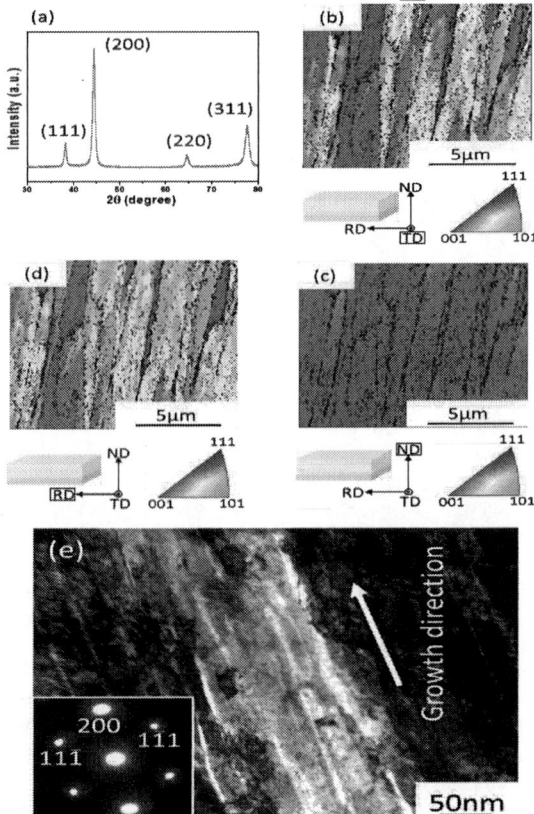

Figure 11. (a) XRD pattern and EBSD mapping coresponds to the (b) TD, (c) ND, and (d) RD, and (e) bright field TEM image of Au film.

$$\sigma_{Y,Pillar\ 2} = \frac{\tau_{CRSS}}{\cos\theta\cos\varphi} \ , \qquad\qquad [4]$$

where, $\cos\theta\cos\varphi$ is the Schmid's factor. The Schmid's factor for single crystals deformed along [100] direction is 0.408. Ratio of the σ_Y's ($\sigma_{Y,Pillar\ 1}/\sigma_{Y,Pillar\ 2}$) obtained from eqs. [3] and [4] is 1.26, which shows the estimated σ_Y of pillar 1 is higher than the σ_Y of pillar 2. This is consistent with the σ_Y's obtained from the stress-strain curves, the σ_Y of pillar 1 is 650MPa and the σ_Y of pillar 2 is 300MPa. On the other hand, ratio of the σ_Y's obtained from stress-strain curves is about 2.16. The difference in the ratio might be attributed to the texture boundary. The texture boundary is expected to affect the strength similar to the grain boundary (37).

Average width of the textures is 1.34µm, and dimensions of the micro-pillar are $10\times10\times20$ µm^3. This implies that pillar 1 would contain more number of textures and of course larger texture boundary area than that in pillar 2. Therefore, the texture boundaries contributed to the strengthening and resulted higher ratio of the σ_Y's.

Early-Stage Diagnosis of Parkinson's Disease

This chapter describes the validity of applying the MEMS accelerometer to the diagnosis of Parkinson's disease.

Background

Analysis of abnormal gait can provide important information about diseases and injuries. For example, patients with Parkinson's disease (PD) often exhibit shuffling, festinating, and freezing of gait. The most widely used clinical rating scale for PD, the Unified Parkinson's Disease Rating Scale, includes observation of gait (38). Patients with cerebellar disorders sometimes have a wide-based (atactic) gait, and those with cerebral vascular disease sometimes exhibit a hemiplegic gait. Recent articles have reported changes in gait, such as reduced gait velocity and stride length, in diseases with gait disorders and in other conditions, such as Alzheimer's disease (39) and depression (40).

Clinical gait analysis is performed mostly by health-care providers using visual observation (41). Although this method is the most readily accessible means of gait analysis available to health-care providers (42), it is a subjective and qualitative method that is inadequate for assessing changes in gait features during ongoing treatment interventions. It is also difficult for clinicians to share this information with health-care providers and patients. Motion capture systems are used in clinical research for gait analysis (43) and scientific research (44). Because they provide well-quantified and accurate results, these systems are currently considered to be the criterion standard for clinical gait analysis (45). While the special equipment needed for motion capture is expensive and requires a large space, few medical institutions can use these systems for clinical gait analysis (42).

Several studies have proposed gait analysis methods using inertial measurement units (IMUs) to solve the problems (46-50). IMUs used in these methods are inexpensive and wearable (51). In particular, we focused on methods that estimate trajectories of a foot because such methods can be used to obtain several spatial gait parameters. Sabatini et al. proposed an IMU-based gait analysis method that estimates a two-dimensional trajectory in the sagittal plane of a foot during walking (46). Other studies have proposed gait analysis methods that estimate the three-dimensional foot trajectory during walking in a stepwise manner to obtain values of foot clearance (48, 50, 52). The trajectory estimation methods reported in several studies (46, 50, 52-54) use an IMU attached on the dorsum of the foot and are better for obtaining this gait feature. As described above, and to the knowledge of the authors, there is no report of a method that estimates three-dimensional foot trajectory from an IMU attached on the shank during waking in a stepwise manner to calculate simultaneously spatial and temporal clinical gait parameters, including stride length, gait speed, stride duration, stance duration and swing duration.

In this study, we demonstrate the validity of the proposed giant analysis method using MEMS accelerometer while comparing the motion capture analysis.

<u>Experimental</u>

Our proposed gait analysis system is illustrated in Fig. 11(a). For gait analysis, we used two IMUs (TSND121, ATR-Promotions, Kyoto, Japan; Fig. 11(b)) with a triaxial accelerometer (±8G range), triaxial gyroscope (±1000° per range), and Android OS tablet (ZenPad10, ASUSTeK Computer Inc., Taipei, Taiwan; Fig. 11(c)). Raw accelerometer and gyroscope signals were sampled at 100Hz (16 bits per sample). The size of the IMU is 37mm×46mm×12 mm and its weight is about 22g. IMUs are attached on the shanks (just above the ankles) with bands as shown in Fig 11(b). The inertial coordinate system used to represent foot orientation and position is also shown in Fig. 11(b). Acceleration and angular velocity data of both the shanks measured during walking are transmitted to the tablet through Bluetooth.

We conducted an experimental evaluation of our proposed method to validate the

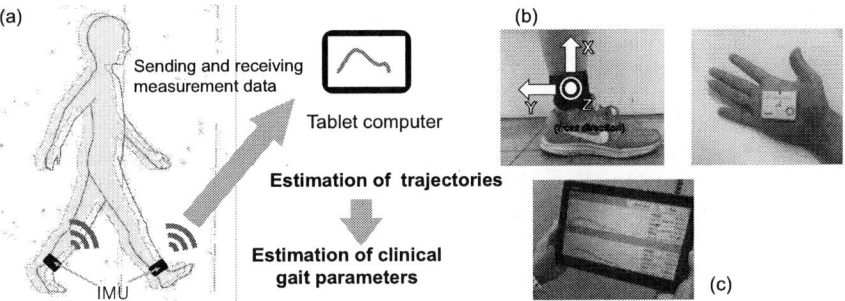

Figure 11. The proposed gait analysis system.

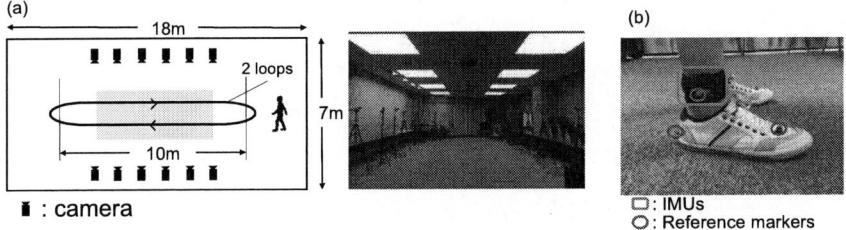

Figure 12. Motion capture system.

accuracy of the trajectory estimation of the shanks (just above the ankles) to verify whether it allows the analysis of gait for clinical purposes. Twenty healthy people participated in the experiment, and we used a motion capture system as the criterion standard for the evaluating gait in a clinical setting. We evaluated the accuracy of our proposed method for calculating the estimated trajectory and clinical gait parameters. We used an IMU attached on the shanks for the experimental evaluation.

An optical motion capture system (Nobby Tech. Ltd., Tokyo, Japan) was used as the reference system. We used 12 cameras, and the motion capture volume was about 2m ±7m ±1m (width, length, and height) as shown in Fig. 12(a). The position error of the markers of the motion capture system during the calibration was less than 1mm. The dimensions of the floor of the room were 18m ±7m (length and width) as shown in Fig. 12(a). Three optical markers were attached on each foot as shown in Fig. 12(b). Two of the three markers were attached on the heel and the toe (metatarsal head II) to assess gait parameters. The third marker was attached on the IMU to evaluate the trajectory estimation. To synchronize the IMU and the motion capture system, the participants hit their heel to the floor before the gait measurement. The peaks of the spike waveforms, that were caused by the heel hits of the participants, in both the IMU signals and the motion capture system data were used to define time 0. We recruited 20 able-bodied volunteer participants, with no history of gait abnormalities, from the Tokyo Institute of Technology for the experimental evaluation. The Ethics Committee of Tokyo Institute of Technology approved the protocols for the evaluation, and all participants provided their written informed consent. Because of technical problems with the motion capture system, the data for one of the 20 volunteer participants were excluded from the analyses. Ultimately, we used data from 19 participants (mean age 23.9 years, 9 men and 10 women, mean height 1.66±0.07m and mean body mass index 20.2±2.7).

As an example of the application of the proposed method to PD patients, we used the method to analyze gait in one healthy elderly participant and four patients with PD. The healthy elderly participant was recruited from a public interest incorporated association in Machida City, Tokyo that provides human resource services for elderly people. Patients with PD were recruited from Kanto Central Hospital, Tokyo. PD had been diagnosed by a physician. The exclusion criteria for this study were past history of other neurological or orthopedic disorders that can affect gait or posture (excluding PD). The healthy elderly participant and the four PD participants provided written informed consent in accordance with requirements of the Ethics Committee of Tokyo Institute of Technology. The Kanto Central Hospital Ethics Committee and the Ethics Committee of Tokyo Institute of Technology approved the protocol for the application of the proposed method.

To construct the spike waveforms for synchronization between the IMU and the motion capture system, the participants hit their heel to the floor before the gait measurement. In two trials, the participants walked on a flat floor at their own self-selected natural pace and a slow pace. In each trial, the participants walk straight in the motion capture volume and turned outside of the motion capture volume. Thus, each trial comprised four straight walks (two round trips; Fig. 12(a)). We used the gait data obtained as the participants walked in the motion capture volume and removed the gait data as the participants turned outside of the motion capture volume.

To consider the validity of the estimation of foot trajectory from IMUs attached on the shanks, we calculated correlations between stride length estimated with the proposed method, measured with a motion capture marker attached on the IMU, and measured with a motion capture marker attached on the heel.

Results & Discussion

To evaluate the proposed method, the shank (just above the ankle) trajectory and clinical gait parameters calculated by our proposed method were compared with those collected by the motion capture system. The comparisons of trajectory information were conducted in the sagittal plane. Five clinical gait parameters of (1) stride length, (2) gait speed, (3) stride duration, (4) stance duration, and (5) swing duration were compared.

The trajectories of our proposed method and the reference data are shown in Fig. 13. The R value between displacement in the direction of forward movement calculated with the proposed method and measured with a marker attached on the IMU was 0.978 as shown in Fig. 14(a). The R value between the maximum vertical displacement calculated with the proposed method and measured with a marker attached on the IMU was 0.925 as shown in Fig. 14(b). The R value between displacement in the direction of forward movement calculated with the marker attached on the IMU and that measured with the marker attached on the heel was 0.994 as shown in Fig. 14(c). Figure 15 shows the agreement between the proposed method and the motion capture system in Bland–Altman plots. The mean±1 SD accuracy of stride length was 0.054±0.031m as shown in Fig. 15(a). The R value between displacement in the direction of forward movement calculated with the proposed method and measured with a marker attached on the heel was 0.978 (Fig. 15(a)). The mean ±1 SD accuracy were as follows: 0.034±0.039m/s for gait speed (Fig. 15(b)); 0.002±0.020s for stride duration (Fig. 15(c); 0.000±0.017s for stance duration (Fig. 15(d)); and 0.002±0.024s for swing duration (Fig. 15(e)). The shank trajectory over 15 steps for each participant is shown in Fig. 16.

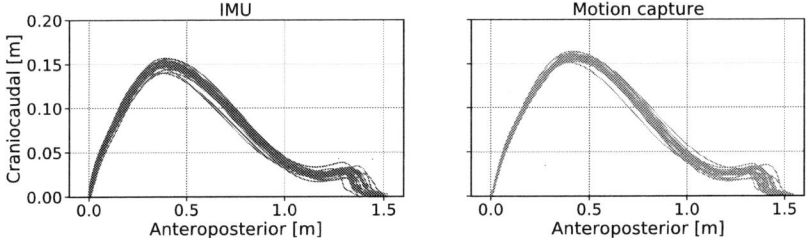

Figure 13. Comparison of Accelerometer and motion capture.

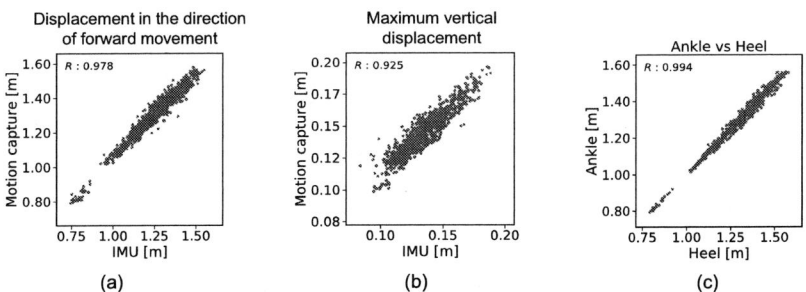

Figure 14. The trajectories of our proposed method and the reference data.

We have proposed a new method for gait analysis that uses IMUs attached on the shanks to estimate foot trajectory and then to obtain estimated clinical gait parameters. The gait parameters obtained with the proposed method consists of stride length, gait speed, stride duration, stance duration, and swing duration. The experimental results show that the proposed method can be used to calculate clinical gait parameters by estimating foot trajectory. The proposed gait analysis method comprises two IMUs with a tri-axial accelerometer, tri-axial gyroscope, and tablet computer. This method can be applied in a variety of locations outside of the gait laboratory and is less expensive than conventional gait analysis methods such as motion capture systems. The clinical advantage is that the patient burden is low because of the light weight (about 24g) and easy attachment of the

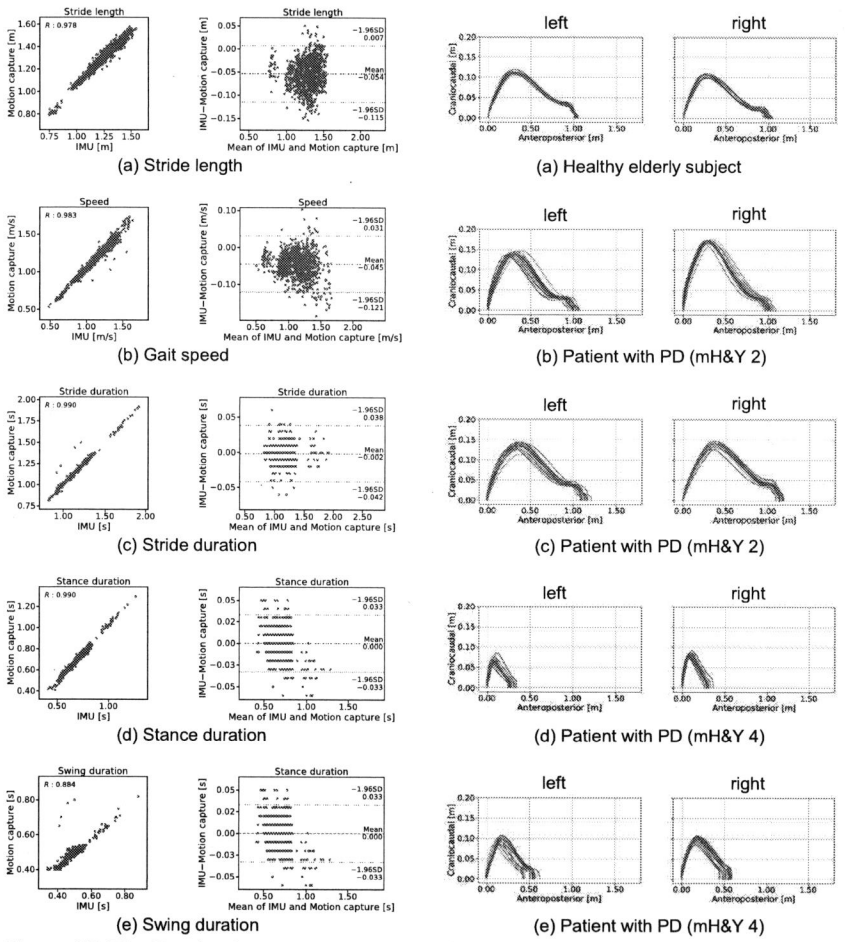

(a) Stride length

(b) Gait speed

(c) Stride duration

(d) Stance duration

(e) Swing duration

Figure 15. The R value between displacement in the direction of forward movement.

(a) Healthy elderly subject

(b) Patient with PD (mH&Y 2)

(c) Patient with PD (mH&Y 2)

(d) Patient with PD (mH&Y 4)

(e) Patient with PD (mH&Y 4)

Figure 16. The proposed method and the motion capture system.

IMUs. We therefore anticipate that the proposed method would be suitable for clinical gait analysis. As for the location of the IMUs, the R value between displacement in the direction of forward movement as measured with the marker attached on the IMU and as measured with the marker attached on the heel (0.994) indicates that the location of the IMUs is valid at least for estimating the stride length. The R value between displacement in the direction of forward movement as estimated by the proposed method and as measured with the marker of the motion capture system attached on the IMU indicates that displacement in the direction of forward movement estimated by the proposed method explained 96% of the variation in displacement in the direction of forward movement as measured with the motion capture system.

The mean error of stride length estimated with the proposed method was 0.054±0.031m. This result suggests that the proposed method can estimate clinical gait parameters such as stride length. A previous method in which the location of IMU is on the dorsum of a foot found that the mean accuracy±precision was 0.015±0.068m (48). The IMU location on the shank may cause bias of this order of accuracy. We expected that further development of the method will overcome this limitation of performance. Several studies (55, 56) have found that stride length is shorter in patients with PD than it is in healthy controls as observed in the example of the application of the proposed method. For example, Morris et al. (57) reported that stride length in PD patients in the off state was 0.96±0.19m, which was shorter than the stride length of 1.46±0.08m as measured in healthy age-matched controls. The R value between displacement in the direction of forward movement estimated by the proposed method and measured with the marker of the motion capture system attached on the heel indicates that stride length estimated by the proposed method explained 96% of the variation measured with the motion capture system. This result suggests that IMUs are potentially useful in clinical gait analysis. We expect further development of the proposed method to evaluate the gait in people with PD.

Summary

We proposed and demonstrated a gold proof-mass MEMS accelerometer for micro-G level sensing. The multi-layer structure of both proof mass and spring can be achieved by the multi-layer metal technology so that B_N can be reduced. The measurement results of the developed MEMS devices were consistent with the design values. The B_N was experimentally obtained to be 22nG/√Hz. The evaluation results confirmed that the proposed MEMS device has potential for micro-G level sensing.

Next, the electroplated Au film was confirmed to be composed of nano-columnar grains embedded in micro-columnar textures. Brittle fracture was observed in the micro-pillar having the long-side perpendicular to the long-axes of the columnar microstructures. Typical multiple slip deformation observed in polycrystalline specimens was observed in the micro-pillar having the long-side parallel to the long-axes. The σ_Y of pillar 1 was roughly two times higher than that of pillar 2. The differences in the deformation behavior and the σ_Y are suggested to be caused by the differences in the compression direction and the total texture boundary area in each pillar.

Finally, our results suggest that the proposed method is suitable for gait analysis whereas there is room for improvement of its accuracy. Unlike methods that use motion capture systems, this method can be used in a variety of locations, such as in the corridor of a medical center. Further development of our proposed method is expected to enable clinicians to share objective information about gait features with health-care providers and patients.

In conclusion, these results of experiments regarding the sensor, the material and the diagnosis suggest that our microgravity sensor can pave the way for the early-stage diagnosis of Parkinson's disease.

Acknowledgments

This work was supported by JST CREST Grant Number JPMJCR1433, Japan.

References

1. D. K. Shaeffer, IEEE Commun. Mag., **51** 100 (2013).
2. J. Laine and D. Mougenot, Proc. TRANSDUCERS 2007 (2007 Int. Solid-State Sensors, Actuators and Microsystems Conf.), 1473 (2007)
3. G. Krishnan, C. U. Kshirsagar, G. K. Ananthasuresh, and N. Bhat, J. Indian Inst. Sci., **87**, 333 (2007).
4. S. Scudero, A. D'Alessandro, L. Greco, and G. Vitale, Proc. 2018 IEEE Int. Conf. on Environmental Engineering (2018), (DOI: 10.1109/EE1.2018.8385252)
5. M. Lemkin and B. E. Boser, IEEE J. Solid-State Circuits, **34**, 456 (1999).
6. B. V. Amini and F. Ayazi, J. Micromech. Microeng., **15**, 2113 (2005).
7. H. Kulah, J. Chae, N. Yazdi, and K. Najafi, IEEE J. Solid-State Circuits, **41**, 352 (2006).
8. B. E. Boser and R. T. Howe, IEEE J. Solid-State Circuits, **31**, 366 (1996).
9. N. Yazdi, K. Najafi, and A. S. Salian, J. Microelectromech. Syst., **12**, 476 (2003).
10. K.-H. Han and Y.-H. Cho, J. Microelectromech. Syst., **12**, 11 (2003).
11. J. Chae, H. Kulah, and K. Najafi, J. Microelectromech. Syst., **14**, 235 (2005).
12. B. V. Amini, R. Abdolvand, and F. Ayazi, IEEE J. Solid-State Circuits, **41**, 2983 (2006).
13. R. Abdolvand, B. V. Amini, and F. Ayazi, J. Microelectromech. Syst., **16**, 1036 (2007).
14. H. Qu, D. Fang, and H. Xie, IEEE Sensors Journal, **8**, 1511 (2008).
15. Y. Dong, P. Zwahlen, A.-M. Nguyen, F. Rudolf, and J.-M. Stauffer, Proc. IEEE/ION Position, Location and Navigation Symposium, 32 (2010).
16. B. Homeijer, D. Lazaroff, D. Milligan, R. Alley, J. Wu, M. Szepesi, B. Bicknell, Z. Zhang, R. G. Walmsley, and P. G. Hartwell, Proc. 2011 IEEE 24th International Conference on Micro Electro Mechanical Systems, 585 (2011).
17. W. Zhu, Y. Zhang, G. Meng, C. S. Wallace, and N. Yazdi, Proc. 2016 IEEE 29th International Conference on Micro Electro Mechanical Systems, 926 (2016).
18. J. Wu, G. K. Fedder, and L. R. Carley, IEEE J. Solid-State Circuits **39**, 722 (2004).
19. S. S. Tan, C. Y. Liu, L. K. Yeh, Y. H. Chiu, M. S.C. Lu, and K. Y. J. Hsu, IEEE Trans. Circuit Syst. I, **58**, 2661 (2011).
20. M.-H. Tsai, Y.-C. Liu, K.-C. Liang, and W. Fang, J. Microelectromech. Syst., **24**, 1916 (2015).

21. T. Konishi, D. Yamane, T. Matsushima, G. Motohashi, K. Kagaya, H. Ito, N. Ishihara, H. Toshiyoshi, K. Machida, and K. Masu, Jpn. J. Appl. Phys., **52**, 06GL04 (2013).
22. D. Yamane, T. Konishi, T. Matsushima, K. Machida, H. Toshiyoshi, and K. Masu, Appl. Phys. Lett., **104**, 074102 (2014) .
23. D. Yamane, T. Konishi, T. Safu, Hiroyuki Ito, H. Toshiyohi, K. Masu, and K. Machida, Proc. Asia-Pacific Conference of Transducers and Micro-Nano Technology (APCOT 2016), Kanazawa, Japan, June 26-29, 299 (2016).
24. D. Yamane, T. Konishi, T. Safu, H. Toshiyoshi, M. Sone, K. Machida, H. Ito, and Kazuya Masu, Sensors and Materials, **31**, 2883 (2019).
25. K. Machida, S. Shigematsu, H. Morimura, Y. Tanabe, N. Sato, N. Shimoyama, T.Kumazaki, K. Kudou, M. Yano, and H. Kyuragi, IEEE Trans. Electron Devices, **48**, 2273 (2001).
26. D. R. Lide, CRC Handbook of Chemistry and Physics (CRC Press, Boca Raton, FL, 1994) 75th ed.
27. W. Yun, R. T. Howe, and P. R. Gray, Proc. Technical Digest IEEE Solid-State Sensor and Actuator Workshop, 126 (1992).
28. J. H. Smith, S. Montague, J. J. Sniegowski, J. R. Murray, and P. J. McWhorter, Proc. International Electron Devices Meeting, 609 (1995).
29. M.-H. Tsai, Y.-C. Liu, and W. Fang, J. Microelectromech. Syst., **21**, 1329 (2012).
30. C. Lu, M. Lemkin, and B. E. Boser, IEEE J. Solid-State Circuits, **30**, 1367 (1995).
31. M. A. Lemkin, B. E. Boser, D. Auslander, and J. H. Smith, Proc. International Solid State Sensors and Actuators Conference (Transducers '97), 1185 (1997).
32. N. Yazdi and K. Najafi, 1999 IEEE International Solid-State Circuits Conference. Digest of Technical Papers. ISSCC. First Edition (Cat. No.99CH36278), 132 (1999).
33. M. Yoshiba, C.-Y. Chen, T.-F.M. Chang, T. Nagoshi, D. Yamane, K. Machida, K. Masu, M. Sone, Mater. Trans., **57**, 1257 (2016).
34. J.R. Greer, W.C. Oliver, W.D. Nix, Acta Mater., **53**, 1821 (2005).
35. G.I. Taylor, J. Inst. Metal, **62**, 307 (1938).
36. E. Schmid, W. Boas, Plasticity of Crystals (Hughes, London, 1950).
37. N. J. Petch, J. Iron, Steel Inst., **174**, 25 (1953).
38. C. G. Goetz, B. C. Tilley, S. R. Shaftman, G.T. Stebbins, S. Fahn, P. Martinez-Martin, W. Poewe, C. Sampaio, M. B. Stern, R. Dodel, B. Dubois, R. Holloway, J. Jankovic, L. Kulisevsky, A. E. Lang, A. Lees, S. Leurgans, P. A. Lewitt, D. Nyenhuis, C. W. Olanow, O. Rascol, A. Schrag, J. A. Teresi, J.J. Van Hilten, N. Lapelle, Movement Disorder Society, U.R.T.F. Mov. Disord., **23**, 2129 (2008).
39. M.M. Mielke, R.O. Roberts, R. Savica, R. Cha, D.I. Drubach, T. Christianson, V. S. Pankratz, Y. E. Geda, M. M. Machulda, R. J. Ivnik, D. S. Knopman, B. F. Boeve, W. A. Rocca, and R. C. Petersen, J. Gerontol. A Biol. Sci. Med. Sci., **68**, 929 (2013).
40. M. R. Lemke, T. Wendorff, B. Mieth, K. Buhl, and M. Linnemann, J. Psychiatr. Res., **34**, 277 (2000).
41. D. E. Krebs, J. E. Edelstein, and S. Fishman, Phys. Ther., **65**, 1027 (1985).
42. S. Barker, R. Craik, W. Freedman, N. Herrmann, and H. Hillstrom, Med. Eng. Phys., **28**, 460 (2006).
43. J. L. Mcginley, R. Baker, R. Wolfe, and M. E. Morris, Gait Posture, **29**, 360 (2009).

44. D. E. Lieberman, M. Venkadesan, W. A. Werbel, A. I. Daoud, S. D'andrea, I. S. Davis, R. O. Mang'eni, and Y. Pitsiladis, Nature, **463**, 531 (2010).
45. M. H. Cameron and J. M. Wagner, Curr. Neurol. Neurosci. Rep., **11**, 507 (2011).
46. M. Sabatini, C. Martelloni, S. Scapellato, and F. Cavallo, IEEE Trans. Biomed. Eng., **52**, 486 (2005).
47. S. T. Moore, H. G. MacDougall, J. M. Gracies, H. S. Cohen, and W. G. Ondo, Gait Posture, **26**, 200 (2007).
48. B. Mariani, C. Hoskovec, S. Rochat, C. Bula, J. Penders, and K. Aminian, J. Biomech., **43**, 2999 (2010).
49. J. R. Rebula, L. V. Ojeda, P. G. Adamczyk, and A. D. Kuo, Gait Posture, **38**, 974 (2013).
50. N. Kitagawa and N. Ogihara, Gait Posture, **45**, 110 (2016).
51. Y. Fujiki, P. Tsiamyrtzis, and I. Pavlidis, CHI'09 Extended Abstracts on Human Factors in Computing Systems, Boston, MA, USA, 3425 (2009).
52. B. Mariani, S. Rochat, C. J. Bula, and K. Aminian, IEEE Trans. Biomed. Eng., **59**, 3162 (2012).
53. Salarian, H. Russmann, F. J. Vingerhoets, C. Dehollain, Y. Blanc, P. R. Burkhard, and K. Aminian, IEEE Trans. Biomed. Eng., **51**, 1434 (2004).
54. Tunca, N. Pehlivan, N. Ak, B. Arnrich, G. Salur, and C. Ersoy, Sensors (Basel), **17**, 825 (2017).
55. H. Stolze, J. P. Kuhtz-Buschbeck, H. Drucke, K. Johnk, M. Illert, and G. Deuschl, J. Neurol. Neurosurg. Psychiatry, **70**, 289 (2001).
56. C. Curtze, J. G. Nutt, P. Carlson-Kuhta, M. Mancini, and F.B. Horak, Mov. Disord., **30**, 1361 (2015).
57. M. Morris, R. Iansek, J. Mcginley, T. Matyas, and F. Huxham, Mov. Disord., **20**, 40 (2005).

Influence of the Biomaterial Thickness in a Dielectrically Charged Modulated Fringing Field Bio-Tunnel-FET Device

C. N. Macambira[1*], P. G. D. Agopian[1,2], J. A. Martino[1]

[1] LSI/PSI/USP, University of Sao Paulo, Sao Paulo, Brazil
[2] UNESP, Sao Paulo State University, Sao Joao da Boa Vista, Brazil
*e-mail: christianmacam@usp.br

In this paper the influence of the biomaterial thickness (t_{Bio}) in a dielectrically charged modulated fringing field Bio-Tunnel FET is studied using TCAD simulation. The sensitivity of the biosensor TFET is observed in the ambipolar current region due to the variation of the dielectric permittivity and/or the presence of charges into the biomaterial. The results show that when t_{Bio} increases from 10 nm to 30 nm, the sensitivity increases up to 2 orders of magnitude. The presence of positive fixed charges into the biomaterial also increases the sensitivity more than 1 order of magnitude in the studied charge magnitude range.

Introduction

The biosensor market is expected to increase more than 40% in 5 years (2019 to 2024) due to the versatility of these devices (1). Then, many research groups are studying new types and optimizations of these devices. One of the promising devices in study is the field effect transistor (FET) that offers many benefits compared to other types of biosensors (2). Among the FETs devices one kind of device has attracted the interest of many research groups. The tunnel field-effect transistor (TFET) is a device that has many advantages compared with conventional FET devices, such as, low power consumption, negligible short channel effects, lower subthreshold slope (3), (4). In this context the use of a dielectrically charged modulated fringing field Bio-Tunnel-FET can contribute to perform fast analysis with a high sensitivity.

Device Characteristics

Figure 1 shows the Bio-TFET cross section which has been simulated with TCAD Sentaurus (5).

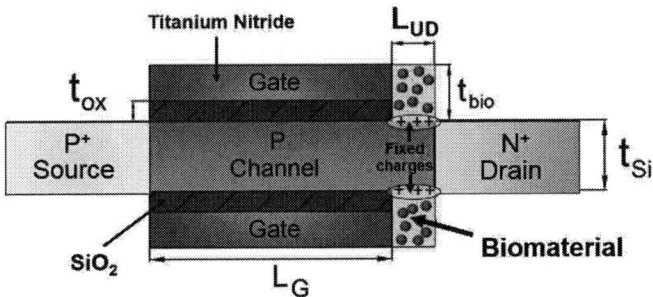

Figure 1. Cross section of dielectrically charged modulated fringing field Bio-TFET device.

The proposed device has a gate length (L_G), equivalent gate oxide thickness (t_{ox}) and channel silicon thickness (t_{Si}) of 40 nm, 1 nm and 10 nm. The drain underlap (L_{UD}) is variable in the range of 1 to 100 nm. The doping profile of the device consists of a source and drain doping concentration of 10^{20} atom/cm^3 of boron and arsenic, respectively. An intrinsic channel with 10^{15} atom/cm^3 of boron. The gate electrode has a workfunction of 4.7 eV. The main models incorporated in the simulation were the bandgap narrowing (BGN), Shockley Read Hall recombination (SRH) and the nonlocal band-to-band tunneling (BTBT). In the drain underlap region is deposited a biomaterial for biosensing purposes. This material has a dielectric constant (k, where $\varepsilon = k*\varepsilon_0$) of 1 (air), 2.1 (Streptavidin), 3.57 (APTES), 8 (Anti-Iris antibody) and 10 (Isoquinoline) (6).

Analysis and Results

Figure 2 shows the Bio-TFET transfer characteristics. The ambipolar region is a parasitic tunneling current that occurs for negative gate bias (V_G) for nTFET at drain to channel region.

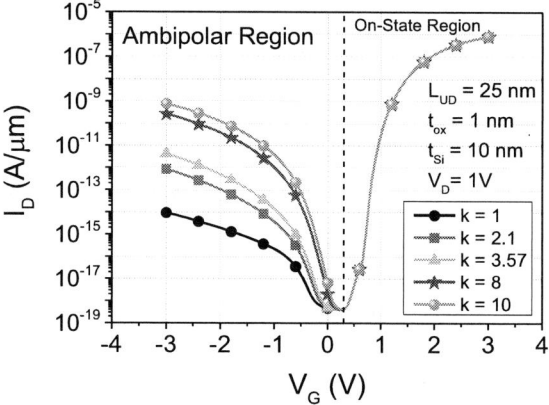

Figure 2. Transfer characteristic curve of the Bio-TFET for different values of k, for a fixed value of L_{UD} = 25 nm and t_{Bio} = 10 nm.

To decreases this effect a drain underlap (L_{UD}) is performed at the device (7). However, when the biomaterial dielectric constant (k) increases, the ambipolar current also increases orders of magnitude, meaning that the device is very susceptible to k variation in the biomaterial region and can be used for biosensing of different types of biomolecules with a distinct k.

In order to quantify the influence of change in k in the biomaterial region the equation in terms of I_D: Sensitivity = $\Delta I_D / I_{D(Ref)}$ was used. Where ΔI_D represents the difference between the drain current of biosensing (output), e.g., the drain current with the presence of biomaterial $I_{D(Bio)}$ and the drain current of reference $I_{D(Ref.)}$ (input), e.g., drain current without biomaterial $I_{D(k=1)}$, both for a fixed value of gate voltage (V_G = -2 V) in the ambipolar region. Fig.3 shows the sensitivity of the device as a function of L_{UD} for different k values.

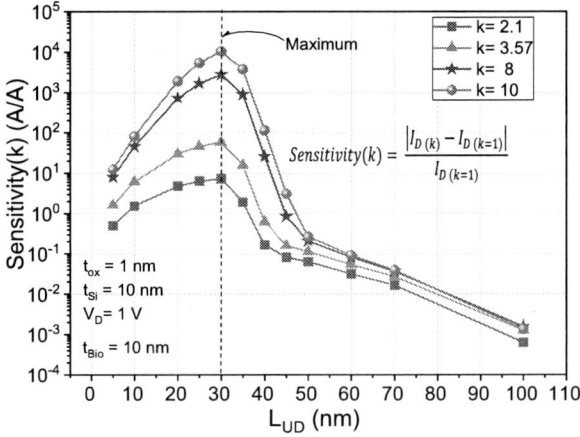

Figure 3. Influence of the underlap length on the sensitivity of the Bio-TFET, for different k values.

The maximum sensitivity occurs for a $L_{UD} = 30$ nm for all range of k values. The sensitivity increases for higher k as reported in (8). Another parameter that is important for sensing is the dimension of the target. Many biomolecules have different shapes, e.g., spherical, ellipsoidal or elongated (9). Therefore, the study of the parameter is fundamental for the functional operation of the device. In Fig.4 the sensitivity is analyzed as a function of biomaterial thickness (t_{Bio}) for a fixed value of $L_{UD}=30$nm.

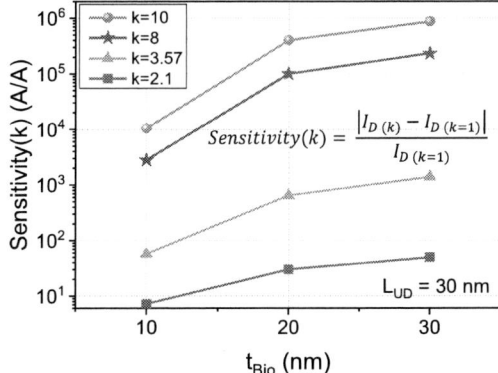

Figure 4. Sensitivity as a function on t_{Bio} length of the Bio-TFET, for different k value and fixed L_{UD} of 30 nm.

With the rising of t_{Bio} and k the sensitivity increases, i.e., exceeding 2 orders of magnitude comparing $t_{Bio} = 10$ nm and $t_{Bio} = 30$ nm for $k = 10$. This increment on sensitivity is due to the tunneling length lowering, as can be noticed in Fig.5.

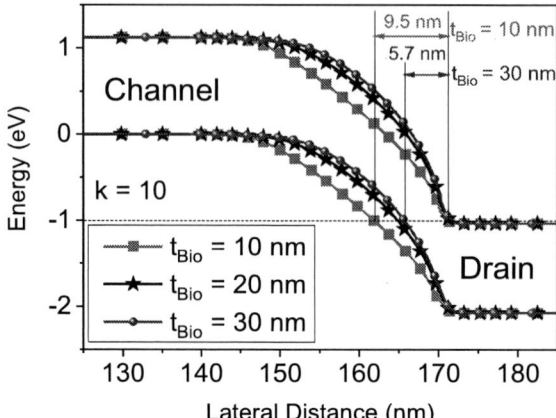

Figure 5. Energy band diagram of the Bio-TFET for different values of t_{Bio}, 1 nm below the interface silicon/oxide in the channel, biased in ambipolar region ($V_G = -2$ V).

The tunneling length decreases from 5.7 nm ($t_{Bio} = 10$ nm) to 9.5 nm ($t_{Bio} = 30$ nm), both for k = 10 and fixed energy of - 1 eV. The lowering of the tunneling length increases the ambipolar current $I_{D(Bio)}$ that enhances the sensitivity of the device. An important consideration for the biosensing process is the charge on the biomaterial that affect the tunneling current and therefore affects the sensitivity of the device. In Fig. 6, the sensitivity is studied as a function of positive fixed charges (Q_{Bio}) in the range of 10^{10} to 10^{12} cm^{-2}. The sensitivity now is in order of the current in the presence of positive fixed charges and without it. The sensitivity increases for thicker values of t_{Bio}. This is due to the higher effect of the fringing field in the underlap region.

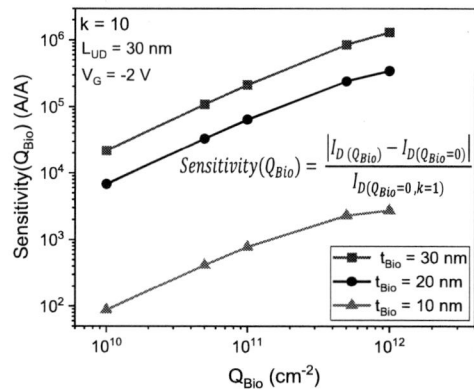

Figure 6. Bio-TFET sensitivity as a function of the biosensor charge density, for different t_{Bio} thicknesses.

Fig.7 shows the absolute electric field of the device. It is possible to notice that the peak electric field corresponds to the channel/drain junction due to negative gate bias applied on the device ($V_G = -2$ V). As Q_{Bio} increases the maximum electric field exhibits a slight decrease. This decrease reflects directly in the sensitivity as observed in Fig.6.

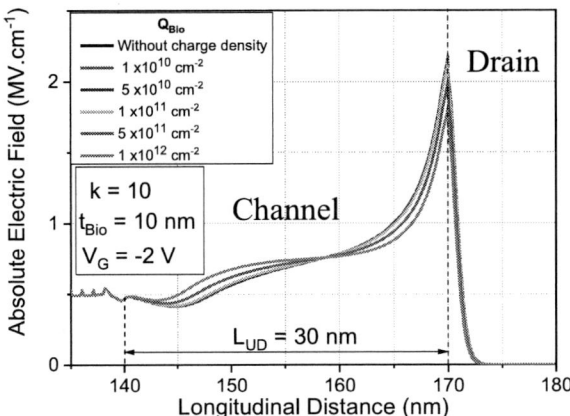

Figure 7. Absolute electric field of the device with t_{bio} = 10 nm in the tunneling region between channel and drain for V_G = -2 V (ambipolar region).

In order to study this effect for different values of t_{Bio} the maximum values of electric field for each Q_{Bio} were used in Fig. 8. As t_{Bio} increases, the effect of the fringing field increases in the biomaterial region and as a consequence, the device becomes more sensitive with the presence of charges.

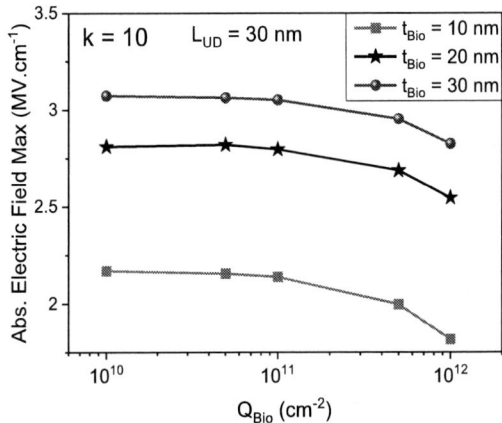

Figure 8. Maximum values of electric field for different interface charge densities and biomaterial thicknesses.

The highest values of maximum electric field were observed for t_{Bio} = 30 nm followed by t_{Bio} = 20 nm and t_{Bio} = 10 nm. The reduction of the electric field is due to the polarization of fixed charges that is positive. For Q_{Bio}=10^{12} cm^{-2} the highest reduction is observed and this point is where the sensitivity has the highest value (Fig.6).

Conclusions

The study of charges and biomaterial thickness in a dielectrically modulated fringing field Bio-TFET was performed in this work. The device shows high sensitivity with the increase of t_{Bio} in the range studied in this work. The highest value of sensitivity(k) was observed for $t_{Bio} = 30$ nm, for all dielectric constants. Furthermore, the effect of positive fixed charges is more notable for thicker t_{Bio}. Then, the t_{Bio} thickness strongly impacts on device sensing.

Acknowledgments

The authors would like to thank FAPESP, CNPq and CAPES for the financial support.

References

1. "Biosensors Market worth $31.5 billion by 2024 Growing with a CAGR of 8.3%",[Online].Available:https://www.marketsandmarkets.com/PressReleases/biosensors.asp, (2019).
2. X. Dai, R. Vo, H. H. Hsu, P. Deng, Y. Zhang, and X. Jiang, *Nano Lett.*, vol. 19, no. 9, pp. 6658–6664, (2019).
3. M. D. V. Martino, J. A. Martino, and P. G. D. Agopian, in *SBMicro 2015 - 30th Symp. Microelectron. Technol. Devices*, pp. 1–5, (2015).
4. N. Guenifi, S. B. Rahi, and T. Ghodbane, *Mater. Focus*, vol. 7, no. 6, pp. 866–872, (2019).
5. "Sentaurus Device User Guide," Mountain View, CA, USA: Synopsys, (2019).
6. R. Goswami and B. Bhowmick, *IEEE Sens. J.*, vol. 19, no. 21, pp. 9600–9609, (2019).
7. P. G. Der Agopian, M. D. V. Martino, S. G. S. Filho, J. A. Martino, R. Rooyackers, D. Leonelli, and C. Claeys, *Solid. State. Electron.*, vol. 78, pp. 141–146, (2012).
8. C. N. Macambira, P. G. D. Agopian, and J. A. Martino, *ECS J. Solid State Sci. Technol.*, vol. 8, no. 3, pp. Q50–Q53, (2019).
9. H. P. Erickson, *Biol. Proced. Online*, vol. 11, no. 1, pp. 32–51, (2009).

Study of a Charge-Based Biosensor and Reconfigurability using [BE]SOI MOSFET

L. S. Yojo[a], R. C. Rangel[a,b], K. R. A. Sasaki[a] and J. A. Martino[a]

[a] LSI/PSI/USP, University of Sao Paulo, Sao Paulo, Brazil
[b] FATEC-SP, Faculdade de Tecnologia de São Paulo, Sao Paulo, Brazil
correspondent author email: leonardo.yojo@usp.br

The Back Enhanced ([BE]) SOI MOSFET is a device whose operation can be tuned by the back gate bias (V_{GB}), i.e., it can act as an n- or p-type MOSFET if V_{GB} is positive or negative enough, respectively. This characteristic was explored in the bio-sensing application, in which a charge-based sensor was studied through simulation. Physical dimensions (gate and underlap lengths, gate oxide, silicon and buried oxide thicknesses) were varied in order to determine the most sensitive case. The buried oxide thickness presented highest variation in the results, showing the importance of the substrate on its operation. The biomaterial charge concentration analysis showed a better sensitivity of the device for positive charges when biased as an n-type transistor, while the p-type bias presented better sensitivity to negative charges. Thus, the versatility of the [BE]SOI MOSFET can be used as an advantage in the field of biosensors.

Introduction

Sensors that are capable of detecting biological components have been attracting attention due to the numerous application possibilities, for instance, the food industry and the point of care health may benefit from the use of biosensors (1). In this context, the use of field effect transistors (FET) based sensors brings the advantages of this technology to the biosensing field such as mass production reduced cost, small size, signal processing circuit integration capability and possibility of higher sensitivity (2).

The back enhanced silicon-on-insulator ([BE]SOI) metal-oxide-semiconductor (MOS) FET (3) was developed and fabricated at Integrated System Laboratory (LSI) of University of Sao Paulo (USP), Brazil. The main attribute of this device is the reconfigurability, i.e., it can operate as an n-type or as a p-type transistor, depending on the back-gate bias. This work explores the characteristics of the [BE]SOI MOSFET as a charge-based biosensor (4) through numerical simulations, using Synopsys Sentaurus TCAD (5), based on experimental measurements.

Device characteristics

Figure 1 shows the schematic profile of the simulated device. It is a planar transistor with a non-intentionally doped silicon channel (intrinsic acceptors dopant concentration of $1x10^{15}$ cm^{-3}). Schottky drain/source contacts were considered (3). The two underlap regions between the drain and gate electrodes and between the source and gate electrodes are the areas where the biological material of interest is deposited, and, depending on its charge concentration (Q_{oxbio}), the drain current conduction can be affected, as shown in figure 2. The operation principle of the device is based on the back-gate bias (V_{GB}). If V_{GB} is positive (n-type ^{BE}SOI), electrons are induced at the back interface of the silicon channel (between the silicon and the buried oxide). The carrier layer at the back interface enables the drain current conduction that can be modulated by the front gate voltage (V_{GF}). A negative enough V_{GF} fully depletes the silicon channel, interrupting the current, whereas a positive enough V_{GF} can induce electrons also at the front interface (between the silicon channel and the gate oxide), decreasing the channel resistance (6). An analogous operation occurs for the p-type ^{BE}SOI, however, a negative V_{GB} is needed and holes are induced at the back interface.

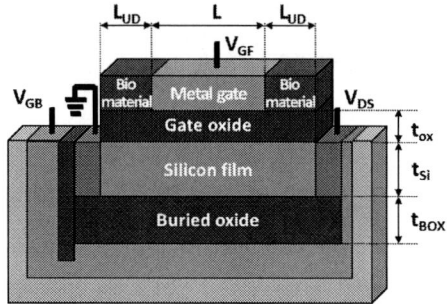

Figure 1. Schematic profile of the ^{BE}SOI MOSFET biosensor.

Simulations results and discussion

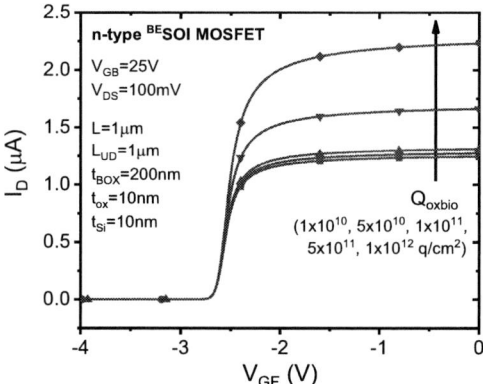

Figure 2. Drain current simulation of the n-type ^{BE}SOI as a function of the biomaterial charge concentration.

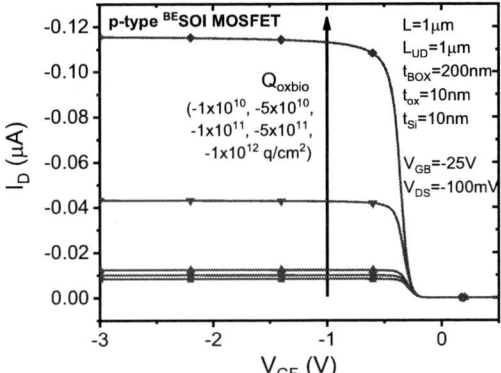

Figure 3. Drain current simulation of the n-type ^{BE}SOI as a function of the biomaterial charge concentration.

In the figure 2 and figure 3, positive Q_{oxbio} values, $V_{GB}=25V$ and $V_{DS}=100mV$ were used in the n-type ^{BE}SOI while the p-type was biased with $V_{GB}=-25V$ and $V_{DS}=-100mV$ and negative Q_{oxbio} values were considered. The drain current increased as a function of $|Q_{oxbio}|$ because the biomaterial charges electric field acts in the sense of increasing the induced carriers in the silicon channel.

In order to evaluate the influence of the charge concentration of the biological material and the transistor dimensions on the drain current, the simulations were performed varying Q_{oxbio} ($1x10^{10}$, $5x10^{10}$, $1x10^{11}$, $5x10^{11}$ and $1x10^{12}$ q/cm^2), the gate electrode length L (0.1, 0.25, 0.5 and 1 µm), the underlap length L_{UD} (0.05, 0.1, 0.2, 0.5 and 1 µm), the buried oxide thickness t_{BOX} (0.1, 0.15, 0.2 and 0.25 µm), the gate oxide thickness t_{ox} (5, 10, 15 and 20 nm) and the silicon film thickness t_{Si} (5, 10, 15 and 20 nm). The relative permittivity of the electrolyte was set to 80 (7) and positive gate oxide effective charges were considered to fit experimental results.

A sensitivity parameter was defined in equation 1 in order to evaluate the influence of the biomaterial charge concentration, where a relative drain current difference is calculated adopting the lowest Q_{oxbio} ($1x10^{10}$ q/cm^2) curve as the reference. A constant absolute front gate overdrive voltage $|V_{GT}=V_{GF}-V_T|$ of 3V was considered in the calculations in order to compare the results of the n-type and p-type biased devices, where V_T is the threshold voltage.

$$S=\Delta I_D/I_{D,ref} \qquad [1]$$

Where $\Delta I_D = I_D-I_{D,ref}$ is the difference between the drain current extracted from the curve of the analyzed Q_{oxbio} (I_D) and the drain current from the reference Q_{oxbio} ($I_{D,ref}$).

Figure 4 and figure 5 show the sensitivity as a function of each dimension for $Q_{oxbio}=1x10^{12}$ q/cm^2. The parameter L presented small influence on the S, since the front gate electrode length only contributes to alter the channel resistance. The underlap influences the S because its length L_{UD} modifies the area in which the biological material charges affect the channel. The buried oxide thickness t_{BOX} influences inversely the S as the thicker buried oxide diminishes the electric field effect in the channel due to V_{GB}. The t_{ox} and t_{Si} parameters presented different trends for the n-type and p-type biases due to the positive gate oxide effective charges. In the n-type case, the gate oxide charges act in favor to induce electrons also at the front interface, thus the biomaterial charges affect the front interface conduction and the thinner t_{ox} and thicker t_{Si} present higher sensitivity. In the p-type bias, the positive gate oxide charges induce electrons at the front interface, suppressing the front interface conduction. The biomaterial charges affect mostly the carriers at the back interface, thus the thicker t_{ox} and thinner t_{Si} present higher sensitivity. In addition, the p-type biased transistor presented higher sensitivity compared to the n-type.

The sensitivity as a function of the Q_{oxbio} is shown in figure 6, in which negative charges were also considered for the n-type biased and positive charges for the p-type biased devices. A higher sensitivity was obtained in the n-type biasing when positive biomaterial charges were simulated. The opposite was observed for the p-type case, in which negative biomaterial charges showed better sensitivity.

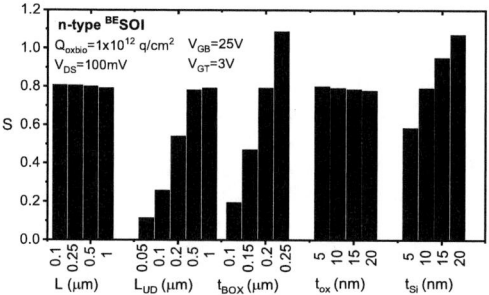

Figure 4. Sensitivity of the n-type ^{BE}SOI charge-based sensor as a function of the parameters. Comparison made adopting the reference device L=1μm, L_{UD}=1μm, t_{BOX}=200nm, t_{ox}=10nm and t_{Si}=10nm for the n-type and p-type biases.

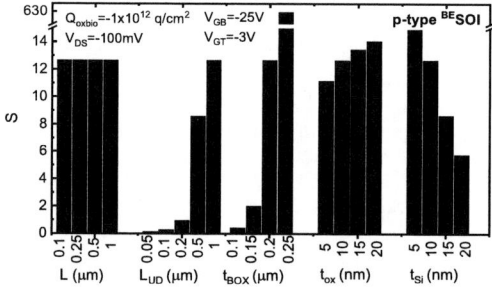

Figure 5. Sensitivity of the p-type ^{BE}SOI charge-based sensor as a function of the parameters. Comparison made adopting the reference device L=1μm, L_{UD}=1μm, t_{BOX}=200nm, t_{ox}=10nm and t_{Si}=10nm for the n-type and p-type biases.

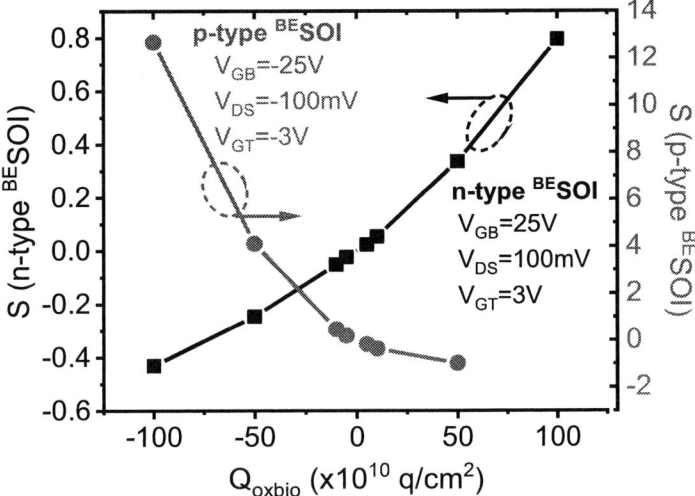

Figure 6. Sensitivity as a function of the biomaterial charge concentration for the n-type and p-type biases (L=1μm, L$_{UD}$=1μm, t$_{BOX}$=200nm, t$_{ox}$=10nm and t$_{Si}$=10nm).

Conclusions

The reconfigurable ^{BE}SOI MOSFET structure was studied as a charge-based biosensor. The underlap regions (between the drain and the gate and between the gate and the source) were explored as sensitive areas, where a biomaterial of interest was deposited in order to observe the effect of its charges on the drain current of the transistor. The simulations varying the physical parameters showed that the underlap regions length and the buried oxide, silicon and gate oxide thickness influence the sensitivity of the sensor.

The fact that the ^{BE}SOI MOSFET is a reconfigurable transistor means that it is possible to obtain a higher sensitivity in a single device if the biological material presents positive or negative charges. N-type biasing can be more suited for detecting positive charges, while p-type biasing may present better sensitivity to negative charges. This characteristic enables a choice of operation, particular to reconfigurable devices, in order to obtain the best sensitivity.

Acknowledgments

The authors would like to thank FAPESP under grant 2018/01568-4, CNPq and CAPES for the financial support.

References

1. R. Ahmad, T. Mahmoudi, M.-S. Ahn and Y.-B. Hahn, *Biosensors and Bioelectronics*, **100** (2018).
2. S. Singh, P. N. Kondekar and N. K. Jaiswal, *Microelectronic Engineering*, **149** (2016).
3. R. C. Rangel and J. A. Martino, *Microelectronics Technology and Devices (SBMicro)*, (2015).
4. L. Yojo, R. C. Rangel, K. R. A. Sasaki and J. A. Martino, *Proc. of SOI-3D-Subthreshold Microelectronics Technology Unified Conference (S3S)* (2019).
5. Synopsys TCAD, Sentaurus Device User Guide, (Version O2018.06).
6. L. Yojo, R. C. Rangel, K. R. A. Sasaki and J. A. Martino, *ECS Trans.*, **85**, 8 (2018).
7. N. Kannan and M. J. Kumar, *IEEE Transactions on Electron Devices*, **62**, 8 (2015).

Study of Underlapped Finfets Behavior for a Radiation Sensing Purpose

W. S. Fonseca and P. G. D. Agopian

Sao Paulo State University (UNESP), Sao Joao da Boa Vista, Brazil
williamfonseca2@hotmail.com

In this paper the electrical characteristics of the Underlapped FinFET transistor is studied for different underlap lengths and underlap oxide material. Disturbs caused by underlap region, such as series resistance effect and the spreading electric field have been evaluated. Besides the underlap length, the impact of oxide permittivity on the on-state current level, fringing fields increasing and on current density was also evaluated. When analyzing the Underlapped FinFET as a radiation sensor, the better sensitivity was obtained for long underlap regions and for transistor biased with high drain voltage. The potential increasing in the underlap region results in on-state current enhancement, which characteristic allows this device present good response for sensing purposes.

Keywords - Underlapped FinFET, Experimental electrical characterization, Radiation sensors.

Introduction

Multiple gate transistors have been used to replace the conventional MOSFETs, mainly due to their better electrical characteristics like higher drain current (50% higher than in a singles-gate MOSFET), improved short channel effects (SCE), smaller drain induced barrier lowering (DIBL), smaller subthreshold slope (SS) (1)(2) and improved power consumption (3). Thanks to these FinFET features, this transistor appears in the Moore's law (4) as an alternative for technological nodes below 32 nm.

In addition, the transistors with an oxide in a gate-to-drain region and/or in a source-to-gate one, called underlapped transistors, have also shown benefits like better DIBL, smaller saturation current degradation due to better control of the charges into the channel (1)(5), high speed performance and also, for analog applications, an improvement in intrinsic dc voltage gain (A_{VO}) and lower cut-off frequency (f_T) (6).

Devices with underlap regions with high k permittivity present an increase of fringe capacitance component of total gate capacitance (C_{GG}) and the extrinsic S/D series resistance generated by the absence of the gate under the channel (7).

Furthermore, the use of MOSFET as radiation sensors has been intensively studied in recent years. Since the main effects caused by the total ionization dose (TID) are well known, the application of this technology as a dosimeter for many applications has greatly increased. One example is the application of MOSFETs as dosimeter for Radiotherapy (8). Knowing that SOI FinFETs have high immunity to soft errors and narrow device channel also presents a good radiation hardness for TID effects, its structure can be applied as an accurate radiation sensor since only the oxide in the underlap region will be responsible for the drain current (I_D) variation.

This paper studies the impact of the underlapped region on electrical characteristics of FinFET transistor, evaluating the behavior of the underlapped length (L_{SP}), as well as the underlap material permittivity constant (k). The transistor sensing ability was also analyzed.

Devices Characteristics and Simulation

From now on, the simulation has been calibrated with experimental self-aligned FinFET behavior for structure with the following characteristics: channel length equal to 400 nm , fin thickness (W_{FIN}=20nm) and fin height (H_{FIN}) of 65nm. The effective gate oxide thickness (T_{OX}) was defined as 2nm and below the channel as 145nm (t_{BOX}). Such oxides will be evaluated following this project. The doping established for S/D was fixed at 10^{20} cm^{-3} with Arsenic atoms whereas in the channel and S/D extensions (N type) the Boron atoms concentration was 10^{15} cm^{-3}.

The simulations were performed with the gradual variation of L_{SP}, from 0 (self-aligned structure) to 50nm, thus generating a gate length (L_G) different from channel length (L_{CH}) as described by equation [1]. In this region were used different materials permittivity (k), simulating the absence of oxide (k=1), silicon oxide (SiO_2 - k=3.9), aluminum oxide (Al_2O_3 - k=9) and hafnium oxide (HfO_2 - k=25). All simulations were performed at room temperature (300K), gate voltage (V_G) from 0 to 1.5 V and 50 mV and 1V for drain bias.

$$L_G = L_{CH} - 2 \times L_{SP} \qquad [1]$$

Figure 1 presents the schematic underlapped FinFET structure.

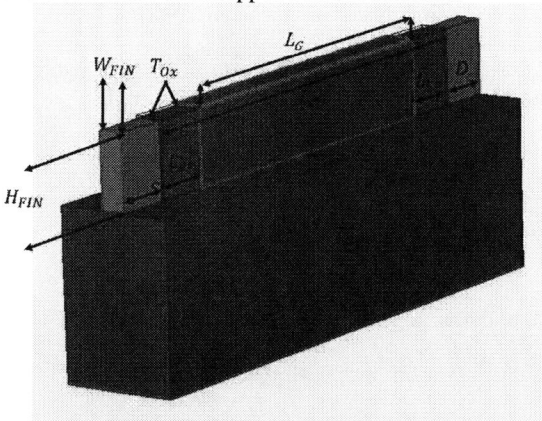

Figure 1. Underlapped FinFET Structure

The current work was performed using the TCAD Sentaurus simulator developed by Synopsys (9). The simulation models take into account the mobility dependence, temperature and doping concentration, the transversal electric field, the generation and recombination phenomena and the high field saturation.

Analysis and Discussion

The experimental behavior of self-aligned FinFET with fin width (W_{FIN}) of 20nm before and after proton irradiation are presented in Figure 2. As already reported in (10) and (11), the FinFET with W_{FIN}=20 nm shows a high immunity to TID effects of proton radiation even considering the accumulated charges in the front and buried oxides due to the high coupling between gates.

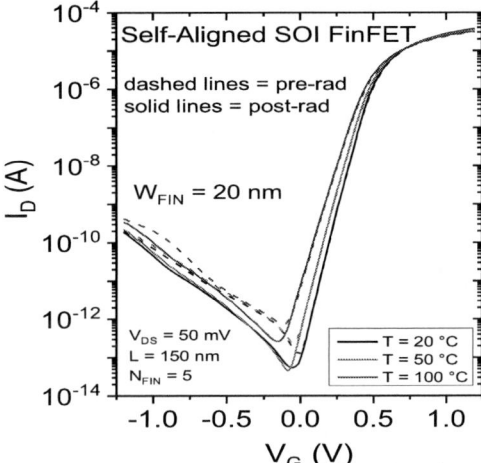

Figure 2 - Experimental drain current as a function of gate bias of FinFETs with W_{FIN}=20nm, pre and post proton irradiation.

Figure 3 presents the drain current (I_D) as a function of gate voltage (V_D) for underlapped FinFETs, varying the underlap length and the permittivity of the oxide in the underlap region. The increase of the L_{SP} results in a strong degradation of I_D due to the lost of the charge control over the underlap, caused by the lower electric field effect (fringing field), which increases the series resistance and the current spreading in the spacer.

Besides the underlap length, the impact of permittivity on the on-state current level (I_{ON}) can also be observed in Figure 3. The permittivity increase results in a higher fringing field from the gate to the source/drain regions and consequently a higher current density that, in turn, causes I_{ON} improvement. The electric field and the electron density were extracted from numerical simulations for devices with k=1 and k=25. A higher fringing field and consequently higher electron density in the underlap region for devices with k=25 can be observed in Figure 4.

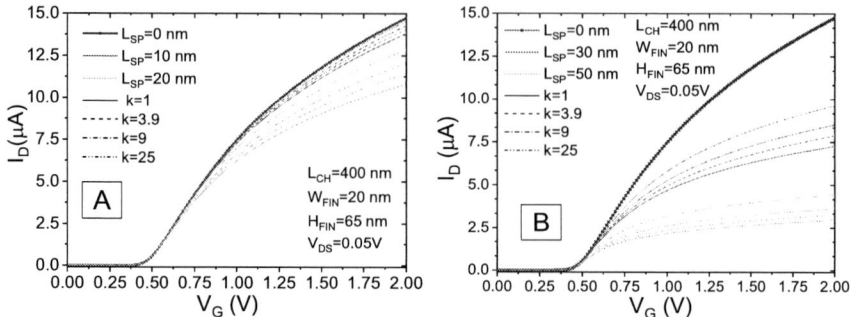

Figure 3. Drain current as a function of gate bias for Underlapped FinFETs for different spacer oxide materials (air, SiO_2, Al_2O_3 and HfO_2) and different underlap lengths 0, 10 e 20 nm(A) and 0, 30 e 50 nm (B).

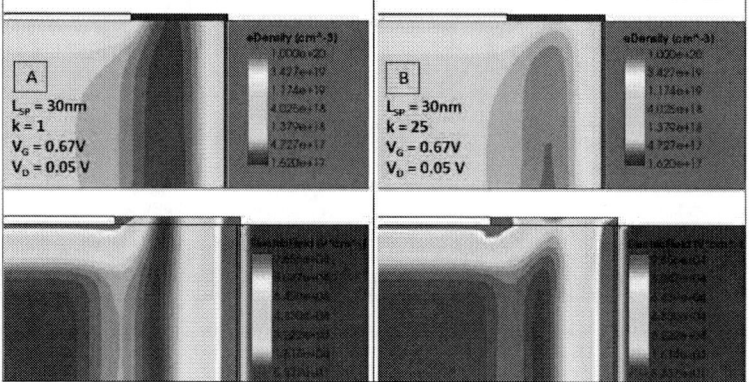

Figure 4. 3D Underlapped FinFET cross-section: Electron density and electric field in the gate to drain underlap region for L_{SP}= 30nm. k =1 (A) and k=25 (B).

Aiming to have a comparison about the self-aligned FinFET behavior and the underlapped counterpart, the sensitivity (S) was calculated by [2]. From Figure 5, it is possible to observe that the greater the L_{SP} and permittivity, the better is the device sensitivity, and it was evaluated in order to select the better oxide and length to the underlap region for further analysis. Although devices with longer underlap presents a higher series resistance and consequently a smaller drain current level, the I_{ON} enhancement caused by the higher influence of fringing field becomes more pronounced when using the higher permittivity, resulting in a more sensitive device. However, for high drain bias (V_{DS}=1V) the underlap length plays a role until LSP=30 nm, for longer underlap no significant sensitivity variation was obtained.

$$I_{D\ (k=x)}/I_{D\ (k=1)} \qquad\qquad\qquad [2]$$

Considering that the device with longer underlap region (L_{SP}=50nm) and higher k (HfO_2–k=25) presents a better sensitivity, it was considered to evaluate its charges sensing ability.

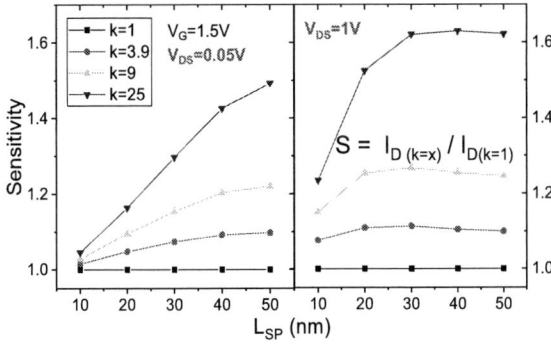

Figure 5. Sensitivity for FinFET with spacers from 0 to 50 nm with the change of the dielectric constant in the spacer region.

Aiming to analyze the underlapped FinFET as a radiation sensor (TID), fixed charges density (Q_{ox}) with carrier concentration of 0, 10^{11}, 5×10^{11} and 10^{12} cm^{-2} were considered in the underlap region, where $Q_{ox} = 0$ cm^{-2} characterizes the pre-radiated transistor. Figure 6 shows that despite the low on-state current caused by the high series resistance, when the oxide charges are considered, there is a potential increase in the underlap region causing an enhancement of I_{ON} (better channel inversion and lower series resistance).

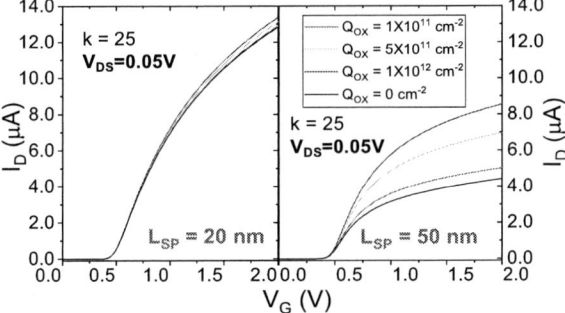

Figure 6. I_D x V_G varying the density of oxide charges for $L_{SP} = 20$ nm (left) and $L_{SP} = 50$ nm (right).

The radiation sensitivity was calculated by post-radiated I_{ON} over pre-radiated I_{ON} ratio (Figure 7).

Although narrow FinFETs are well known as a radiation hardness transistor, through the analysis was possible to notice that a long underlap region with a high k dielectric allows this device to have a good sensing behavior. When considering a relatively low trapped charge density ($Q_{ox} = 5 \times 10^{11}$ cm^{-2}), the drain current increases about 60%, i.e., sensitivity equal 1.6. Increasing the trapped charge density to $Q_{ox} = 1 \times 10^{12}$ cm^{-2}, the obtained sensitivity values are still higher, reaching an increment of 140% on the I_{ON} current level (S=2.4), confirming the excellent characteristic of the device working as a radiation sensor.

An exponential growth of transistor sensitivity was obtained, for high charge density, as the L_{SP} is increased. It's clearly notable that the Underlapped FinFET with $L_{SP} = 50$ nm and k=25 (HfO$_2$) has an excellent behavior to apply it in radiation sensors.

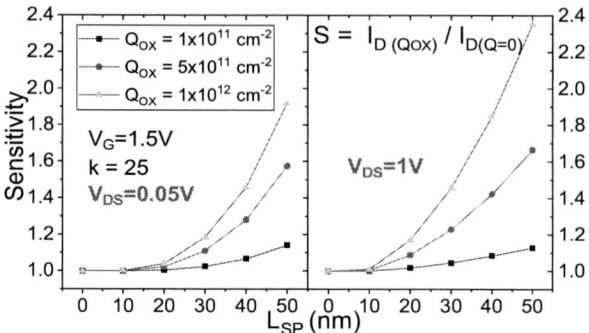

Figure 7. Sensitivity for Underlapped FinFET for L_{SP} from 0 to 50 nm, k=25 and for different charge density in the underlap region.

Conclusions

The impact of the underlap region and the spacer oxide material on the on-state current of FinFETs was evaluated. The absence of gate electrode over channel (underlap region) causes a series resistance increase, that in turns results in a degradation of on-state current. Using this behavior to a sensing purpose, the longer underlap causes a higher sensitivity due to stronger I_{ON} variation. Focusing on the impact of oxide permittivity, a better response was obtained when using materials with high k dielectric, which allows that the electric field lines from the gate strongly affect the channel region under the underlap, increasing the carrier density and consequently reducing the series resistance, enhancing I_{ON} and the device sensitivity.

This work shows that the underlapped FinFET with longer underlap length and high-k spacer oxide might be used as radiation sensor due the increase of sensitivity with the increase of accumulated charge.

References

1. G. Saini and A. K. Rana, *International Journal of VLSI Design & Communication Systems (VLSICS)*, p. 26, (2011).
2. J.-P. Colinge, Silicon-on-Insulator Technology: Material to VLSI, 3rd ed., 255 (2004).
3. M. S. Badran, H. H. Issa, S. M. Eisa and H. F. Ragai, *IEEE Access*, 17256 (2017).
4. E. Mollick," *IEEE Annals of the History of Computing*, 62 (2006).
5. V. Trivedi, J. G. Fossum and M. M. Chowdhury, *IEEE Transaction on Electron Devices*, 56 (2005).
6. A. Kranti and G. A. Armstrong, *IEEE International SOI Conference*, 33 (2007).
7. A. B. Sachid and et al, *IEEE Electron Device Letters*, 129 (2008).
8. M. Garcia-Inza, S. H. Carbonetto, J. Lipovetzky and A. Faigon, *IEEE Transactions on Nuclear Science*, 1 (2016).
9. "TCAD Sentaurus™ Tutorial," Sentaurus, Available: http://www.sentaurus.dsod.pl/. [Accessed 20 10 2019].
10. P. G. D. Agopian, J. A. Martino, D. Kobayashi, E. Simoen and C. Claeys, *IEEE Transactions on Nuclear Science*, p. 707, (2012).
11. L. F. V. Caparroz, C. C. M. Bordallo, J. A. Martino, E. Simoen, C. Claeys and P. G. D. Agopian, *Semiconductor Science Technology*, **33**, 065003, (2018).

CHAPTER 3

Process Technology

Low Temperature SmartCut[TM] process for 3D Integration

W. Schwarzenbach[a], S. Reboh[b], A. Ghorbel[a], G. Gaudin[a], G. Besnard[a], F. Mazen[b], V. Loup[b], S. Maitrejean[b], L. Brunet[b], B.-Y. Nguyen[a] and C. Maleville[a]

[a] SOITEC, Parc Technologique des Fontaines – F-38190 Bernin – France
[b] CEA LETI, Minatec Campus – F-38054 Grenoble – France
Correspondent author email: walter.schwarzenbach@soitec.com

> To support 3D Sequential integration with a cost-effective layer transfer, SmartCut™ process at low temperature (below 500 °C) is proposed. Excellent SOI & BOX layer thickness uniformities are demonstrated, while layer integrity, electron and hole mobility performances are already compliant with development grade requirements.

SmartCut Proposal for 3D Sequential Integration

In support of the semiconductor evolution, 3D integration is considered, addressing several challenges: integration of multiple types of devices including specialized components, continuous footprint reduction as well as optimization of performance and cost (1). However, a key challenge for 3D integration, shown on Figure 1 reported in reference (2), is the preparation of a top tier device using as handle surface a fully processed bottom tier device. Hence, the top tier device thermal budget needs to be reduced to avoid degradation of the bottom devices (3). A mandatory 500 °C thermal budget limitation for 3D Sequential Top layer transfer and MOSFET fabrication is then considered (2, 4).

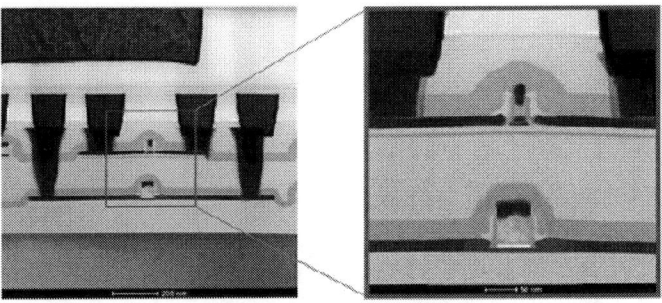

Figure 1: TEM cross section of a 3D structure showing stacked top & bottom tier devices with nanometric alignment – published in (2).

In order to take advantage of the SmartCut process to transfer at reduced cost (i.e. avoiding a conventional but costly SOI bonding & grinding process) a high quality, highly uniform crystalline layer, we propose the use of a new Low Temperature SmartCut process as an alternative for 3D integration. This paper reports on our findings from such transfer demonstrations on bare handle materials, including for the first time SOI-like layer electrical characterization results.

Low Temperature SmartCut for 3D Integration

For 3D integration, SmartCut process is adapted as schematically shown on Figure 2 in order to transfer at low temperature ultra-thin layers compatible with fully depleted requirements. This process option includes the use of engineered, epitaxial processed donor wafers. The epitaxial process on conventional bulk material allows creation of a first Silicon-Germanium layer, to be used as an etch stop layer after SmartCut splitting. Then a subsequent Si layer, defined accordingly to the SOI film thickness target, is epitaxied. Implant, surface preparation, bonding, splitting process steps, all being compliant with maximum 500 °C low temperature requirement, benefit from conventional FD-SOI experience. Finishing process includes selective etching to recover SOI layer (5). Then a dedicated curing process step is considered in order to optimize SOI layer electrical properties. Solid Phase Epitaxial Regrowth (SPER) as described in reference (6) is proposed.

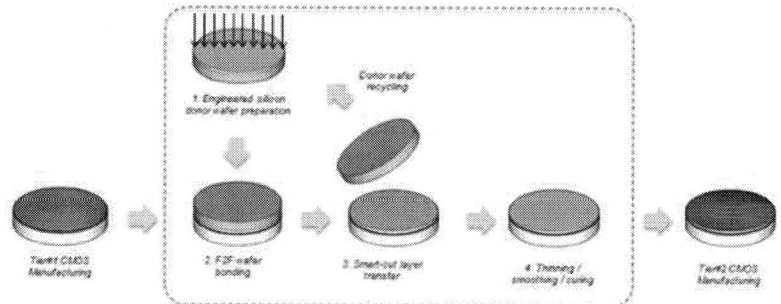

Figure 2: SmartCut Process Flow Adapted for 3D Integration.

Figure 3 highlights such low temperature layer transfer process on a patterned handle wafer. Picture is obtained after SmartCut splitting process step. A visual defect free layer is obtained.

Figure 3: 300mm wafer, Low Temperature Layer Transfer on patterned handle wafer and transferred SOI thickness mapping.

Low Temperature SOI Performance on Bare Handle wafer

The SOI layer thickness variability, over the full spatial wavelength spectrum from device to wafer scale, is driven by the donor epitaxy performance. Figure 4 shows Atomic Force Microscopy (AFM) roughness performance, through 30x30 µm² scan. RMS of 1.0 and 0.9 Å are measured on high and low temperature process options respectively. Best in class performance is then obtained thanks to Low Temperature SmartCut.

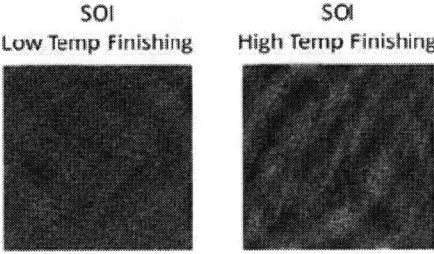

Figure 4: AFM meas., (a) 30x30 µm² scan, (b) PSD graph, 12nm SOI / 20nm BOX, with Low or High Temperature finishing.

Figure 5 shows (a) BOX and (b) SOI layers ellipsometry thickness mapping, on a 12 nm SOI / 20 nm BOX sample. BOX layer variability is primarily driven by thermal oxide growth, thus demonstrating a within wafer uniformity << 10 A. SOI layer thickness variability is mainly driven by epitaxy performance on the engineered donor wafer.

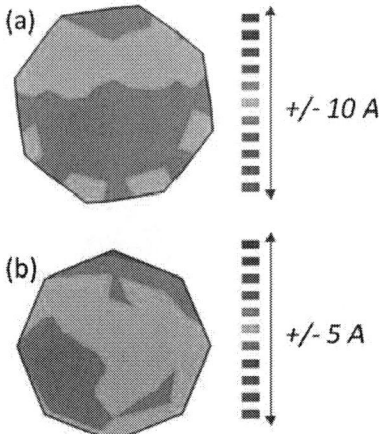

Figure 5: 41 ellipsometry thickness mapping (a) 20nm thick BOX, (b) 12nm thick SOI normalized vs Epi profile, 12/20 Low Temperature sample.

Figure 6 shows typical KLA Tencor SP3 @ 65 nm threshold defect mapping, as measured on 12 / 20 nm low temperature R&D sample. Micro-defect density on this early development product already demonstrates less than 0.1 defect / cm² performance.

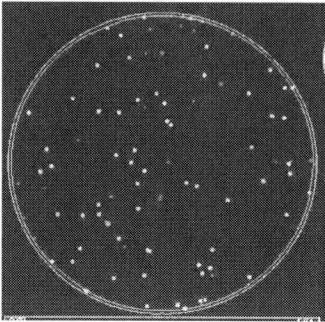

Figure 6: KLA Tencor SP3 @ 65nm threshold surface defect mapping, 12/20nm SOI with Low Temperature process.

Then Figure 7 compares PseudoMOS measured electron and hole mobility in the SOI layer using High and Low Temperature SmartCut processes. A as low as 10% mobility performance difference between both processes vintage is already demonstrated.

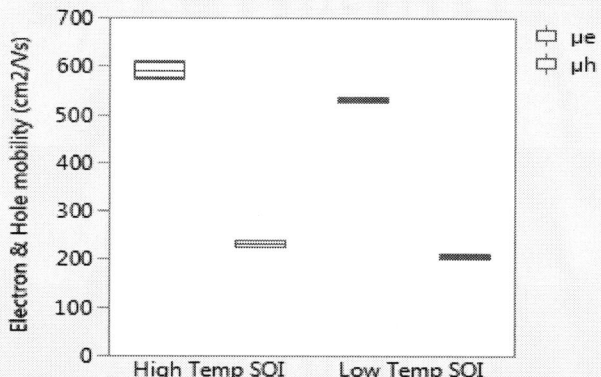

Figure 7: PSiMOS SOI electron (blue) & hole (red) mobility comparison, high vs low temperature Smartcut.

Conclusion

To support 3D sequential integration, layer transfer at low temperature (< 500 °C) is demonstrated thanks to an adapted SmartCut process. Next research step will confirm this capability with low temperature layer transfer on patterned substrate for 3DS device assessment.

References

1. J. Macri, « *Left, Right or Up. Living in a 3D World* », Proc. IEEE S3S Conference, 2018

2. L. Brunet, C. Fenouillet-Beranger, P Batude, S. Beaurepaire, F. Ponthenier, N Rambal, V. Mazzocchi, J.-B. Pin, P. Acosta-Alba, S. Kerdiles, P. Besson, H. Fontaine, T. Lardin, F. Fournel, V. Larrey, F. Mazen, V. Balan, C. Morales, C. Guerin, V. Jousseaume, X. Federspiel, D. Ney, X. Garros, A. Roman, D. Scevola, P. Perreau, F. Kouemeni-Tchouake, L. Arnaud, C. Scibetta, S. Chevalliez, F. Aussenac, J. Aubin, S. Reboh, F. Andrieu, S. Maitrejan, M. Vinet, « *Breakthroughs in 3D Sequential technology* », Technical Digest – IEDM Conference, 2018, pp 7.2.1 – 7.2.4

3. A. Vandooren, L. Witters, J. Franco, A. Mallik, B. Parvais, Z. Wu, W. Li, E. Rosseel, A. Hikkavyy, L. Peng, , N. Rassoul, G. Jamieson, F. Inoue, G. Verbinnen, K. Devriendt, L. Teugels, N. Heylen, E. Vecchio, T. Zheng, N. Waldron, J. Boemmels, De Heyn, D. Mocuta, J. Ryckaert, N. Collaert, « *Key Challenges and Opportunities for 3D Sequential Integration* », Proc. IEEE S3S Conference, 2018

4. C. Fenouillet-Beranger, B. Mathieu, B. Previtali, M.-P. Samson, N. Rambal, V. Benevent, S. Kerdiles, J.-P. Barnes, D. Barge, P Besson, R. Kachtouli, M Cassé, X. Garros, A. Laurent, F. Nemouchi, K. Huet, I. Toqué-Trésonne, D. Lafond, H. Dansas, F. Aussenac, G. Druais, P. Perreau, E. Richard, S. Chhun, E. Petitprez, N. Guillot, F. Deprat, L. Pasini, L. Brunet, V. Lu, « *New insights on bottom layer thermal stability and laser annealing promises for high performance 3D VLSI* », Proc. IEDM Conf., 2014

5. J. Widiez, J.-M. Hartmann, F. Mazen, Y. Bogumilowicz, E. Augendre, C. Veytizou, S. Sollier, M. Martin, M-C. Roure, V. Loup, C. Euvrad, A. Seignard, T. Baron, R. Cipro, F. Bassani, A.-M. Papon, C. Guedj, I. Huyet, M. Rivoire, P. Besson, C. Figuet, W. Schwarzenbach, D. Delprat, T. Signamarcheix, « *SOI-type Bonded Structures for Advanced Technology Nodes* », ECS Trans., 2014, vol. 64(5), pp. 35-48

6. G. Gaudin, W. van den Daele, N. Chartrain, G. Riou, C. Veytizou, I. Radu, S. Cristoloveanu, « *Physical and Electrical Properties of Thin Doped Silicon Films Obtained by Low Temperature Smart Cut and Solid Phase Epitaxy* », ECS J. Solid State Science and Technology, 2013, 2 (12), pp.534-538

Dielectric Science on Today's Devices

D. Misra

Department of Electrical and Computer Engineering, New Jersey Institute of Technology, Newark, NJ 07102, USA

The trends of various high-k gate dielectrics deposited on silicon or germanium is reviewed for logic and memory devices. Initially, Zr and low percentage of Al incorporation into HfO_2-based high-k dielectrics gate stack on silicon through different process conditions were outlined. Subsequently, high-k gate stack deposition process on Ge was evaluated to form a stable Ge/high-k interface. We have also looked at the high-k gate dielectric stacks in a MIM capacitor for possible applications in memory and AI hardware.

Introduction

Device scaling in CMOS technology has approached to 10nm range and below. The gate dielectric, being the critical constraint, has evolved significantly and requires constant quality improvements in order to keep the proper functioning of the transistors and memory devices. While the transistor has transformed from a planar device to a three-dimensional device to a gate all around device, several new devices such as ferroelectric FETs and negative capacitance FETs have emerged to be integrated into standard CMOS technology. Additionally, many forms of memory devices such as resistive random-access memory (RRAM) devices and ferroelectric RAM devices are being investigated for possible implementation of artificial intelligence (AI) hardware. All these devices use high dielectric constant (high-k) materials in some form or other. The electrical performance in these devices depends on the dielectric deposition process, precise selection of deposition parameters, pre-deposition surface treatments and subsequent thermal budget.

For logic devices, once the thickness of SiO_2 reached the onset of direct tunneling region (<1.5 nm) high-k insulators were introduced to suppress the direct-tunneling leakage current. Because of thermal stability, HfO_2-based high-k dielectrics were selected to replace SiO_2. To resolve issues like mobility degradation, threshold voltage control and Fermi-level pinning, HfO_2-based high-k gate dielectrics was introduced with an interfacial SiO_2 layer on silicon devices. Subsequently, higher-k dielectrics are required for transistors performance enhancement to keep up with the scaling without reducing the physical dielectric thickness. It was achieved by controlling the structural phase of HfO_2 by doping HfO_2 with various metal oxides including ZrO_2 as cubic, or tetragonal phase that provides higher-k than amorphous or monoclinic phase. By considering the interface reaction kinetics and thermal budget various atomic layer deposition (ALD) schemes were identified to get a stable dielectric with a robust interface.

In RRAM devices, the resistance change, achieved in a dielectric material between two metal terminals by applying a voltage, is promising for further scaling of memory devices. The switching mechanism in transition metal oxides like HfO_2 is, while evolving, based on similar mechanism to soft breakdown, where a conducting filament path is formed due to oxygen vacancy transition/formation. The process is reversed by rupturing the filament. Various dielectric stacks are currently being studied for RRAM applications. Additionally, ferroelectric characteristics of HfZrO films also derived significant attention recently.

In this work we describe the evolution of dielectric science in nanoelectronics. We discuss the deposition of $Hf_{1-x}Zr_xO_2$ or $HfAlO_x$ on both silicon and/or germanium, a high mobility channel material. The processing conditions to improve the quality of the dielectrics and their unique properties is also discussed. The reliability characterization of these dialectics based on dielectric-semiconductor interface will also be discussed. Some of the current trends in memory devices involving high-k dielectrics will be discussed.

Dielectrics for Logic Devices

High-k Gate Dielectrics on Silicon

Continuous device scaling leads to a decrease in cost per function of technology and improves the economic productivity and the quality of life through proliferation of computers, communication, and other industrial and consumer electronics. Hafnium-based high-k dielectric materials have been successfully used in the industry as a key replacement for SiO_2 based gate dielectrics in high performance logic devices and low standby power logic families in order to continue CMOS device scaling (1). It was because the chip level gate leakage current in a chip level became an issue because of direct-tunneling current with SiO_2 when its thickness approached 1.2 nm. With high-k dielectrics the gate capacitance can be scaled down to meet the scaling requirements without reducing the physical dielectric thickness. Metal gate/high-κ gate stacks allowed the equivalent oxide thickness (EOT) scaling to 0.7 nm or below while keeping gate leakage current within a tolerable limit (2-3).

Initially, the devices with high-k dielectrics faced the challenges of high defect density and it was required to have improved channel mobility, reduced gate leakage current, good control on threshold voltage, lower interface state density and good reliability [2-3]. The HfO_2/Si interface was found to be worse than SiO_2/Si interface and the channel mobility was degraded because of poor interface. Remote phonon scattering became a dominant mechanism in addition to Coulomb scattering. An interfacial layer of SiO_2 was introduced between high-k and silicon to improve the interface quality. Also, high-k deposition methods such as atomic layer deposition (ALD) was introduced to improve the high-k film properties.

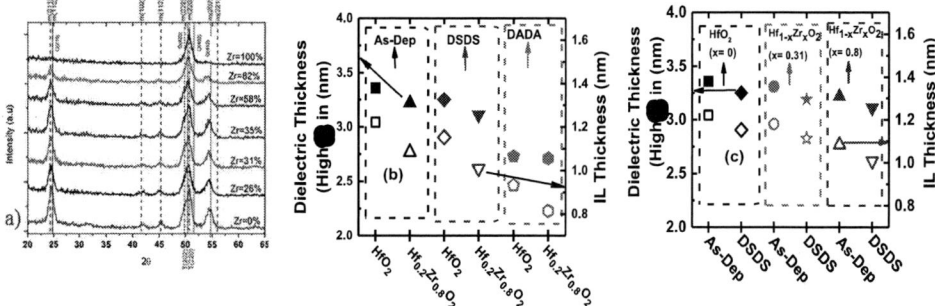

Figure 1 (a) GIIXRD spectra of DADA $Hf_{1-x}Zr_xO_2$ as a function of Zr% in the dielectrics (5), (b) Dielectric thickness (filled symbols on left scale) and interfacial layer (IL) thickness (open symbols on right scale) for MOSCAPs with DSDS, DADA, and As-Dep HfO_2 and $Hf_{0.2}Zr_{0.8}O_2$, (c) Dielectric thickness (filled symbols on left scale) and interfacial layer (IL) for $Hf_{1-x}Zr_xO_2$ with x=0, 0.31, and 0.8 for DSDS and As-Dep processing conditions (6).

The flexibility in ALD technique such as interleaved treatments during the cyclic deposition process of Hf-based high-k dielectrics allowed to improve the dielectric quality. Addition of Zr and Al in HfO_2 enhanced the EOT downscaling and reduced gate leakage current when the interleaved treatments are applied instead of a post deposition anneal (PDA). ALD also allowed Different crystalline phases of HfO_2 and ZrO_2 offers unique characteristics. For example, tetragonal phase offers higher dielectric constant as compared to monoclinic phase which is the thermodynamically stable phase for these dielectrics [3, 4]. In one such experiment when Zr was added to HfO_2 while a cyclic deposition and anneal (DADA) he stabilization into tetragonal phase with a (111) orientation was observed for DADA $Hf_{0.2}Zr_{0.8}O_2$ whereas DADA HfO_2 showed monoclinic crystalline structure (5) as shown in Fig. 1(a). When the cyclic deposition and plasma exposure (DSDS) was carried out during deposition the physical thickness decreased (Fig. 1(b-c)) for different Zr concentration (6) as evaluated by using a MOS capacitor. The leakage current was improved (Fig. 2(a)) due to both Zr addition and plasma exposure as shown in leakage current density J_g as a function of Zr content at a E_{OX} =10 MV/cm in the positive bias region which is mostly influenced by the quality of the interface (6). In addition, the reliability, Weibull plots of time to breakdown (Fig. 2(b)), of these devices improved for higher Zr concentration due to significant reduction of traps in the dielectric as well as improvement of the interfacial layer quality (6).

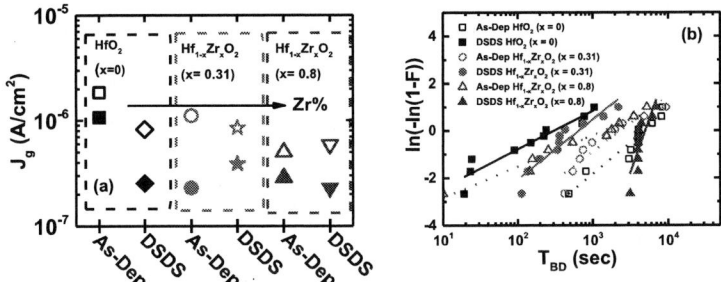

Figure 2. (a) Gate leakage current density for unstressed devices (closed symbols) and for stressed devices (open symbols) for DSDS and As-Deposited MOSCAPs with $Hf_{1-x}Zr_xO_2$ at a constant field stress of E =27.5 MV/cm for 1000s in the gate injection mode, (b) Weibull plot of time to breakdown (T_{BD}) for DSDS and As-Dep $Hf_{1-x}Zr_xO_2$ with different Zr percentages.

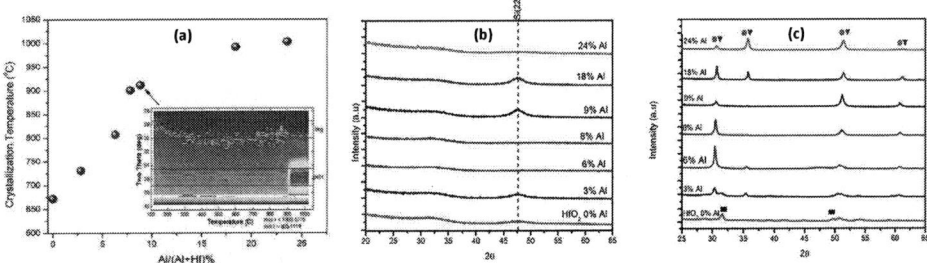

Figure 3. (a) Crystallization temperature of $Hf_{1-x}Al_xO_y$ as a function of Al/(Al+Hf)% in the dielectrics (Inset shows on situ XRD plot at the onset of crystallization for $Hf_{1-x}Al_xO_y$ with x= 0.09); Synchronous grazing in plane X ray diffraction pattern for (b) as deposited ALD $Hf_{1-x}Al_xO_y$, (c) annealed ALD $Hf_{1-x}Al_xO_y$ as a function of Al/(Al+Hf)% (7).

Al Incorporation

Al incorporation used to improve the thermal stability of high-k HfO_2 films (7) as the transition temperature from amorphous to polycrystalline state of $Hf_{1-x}Al_xO_y$ increases. Fig. 3(a) shows the crystallization temperature of $Hf_{1-x}Al_xO_y$ as a function of Al/(Al+Hf)% in the dielectrics (7) where the inset shows the in-situ XRD plot at the onset of crystallization for $Hf_{1-x}Al_xO_y$ with x= 0.09. Fig. 3(b) and 3(c) shows the synchronous grazing in plane X-ray diffraction pattern for as deposited ALD $Hf_{1-x}Al_xO_y$ and annealed ALD $Hf_{1-x}Al_xO_y$ as a function of Al/(Al+Hf)% respectively (7). After annealing the peak position shift toward larger 2θ values in the XRD pattern (Fig. 3(c)) in $Hf_{1-x}Al_xO_y$ due to tetragonal crystalline phase formation (7) whereas monoclinic phase is prominent in as-deposited films (Fig, 3(b)). Annealing, therefore, increases the dielectric constant because of stable tetragonal phase. Incorporating an extremely low percentage of Al (1-7%) in HfO_2 and subsequently annealing (680 °C to 800 °C) shows a decrease in leakage current (J_g) and interface state density (D_{it}) with increase in Al concentration as shown in Fig. 4(a) and Fig. 4(b) respectively (8). Incorporation of higher concentration of Al could exponentially reduce the J_g by improving the quality of the oxide (lower EOT or higher dielectric constant), as Al increased the crystallized temperature. Exposing the dielectric to slot plane antenna N_2 (SPAN) plasma can further reduce the leakage current (Fig. 4(a)). It can be further observed that the D_{it} values were reduced as long as the dielectric remains amorphous (9, 10).

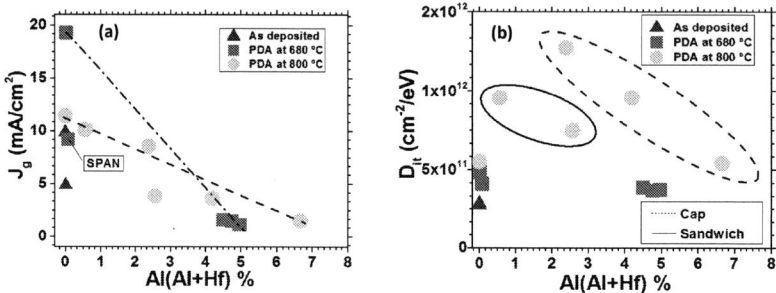

Figure 4. (a) The leakage current (J_g) measured at the gate voltage (V_{FB}-1 V) as a function of the Al/(Al+Hf) % is shown as a function of Al concentration. Dashed dot line and dashed line follow the trend of J_g varies with different Al doping level at 680 °C PDA and 800 °C PDA respectively (SPAN is Slot Plane Antenna in N_2 environment) ; (b) The interface defects density (D_{it}) as a function of the Al/(Al+Hf) % with data points within the dashed line are the samples where Al was introduced by HfAlO as the cap layer on HfO_2 and data points in solid line are the samples where Al was introduced by sandwiching HfAlO inside the HfO_2 stack.

High-k Dielectrics on Germanium

Germanium has been attracting significant interest as an alternative channel material for sub-10 nm technology node to replace silicon because of high carrier mobility. It is possible because of the successful integration of high-ka gate dielectric in silicon technology. Establishing a high-quality interface between the semiconductor and high-k dielectric with reduced interface state density was required to achieve high performance and low power devices. Interface passivation and interface treatments such as wet chemical oxidation, thermal oxidation, electron cyclotron plasma resonance oxidation, ozone ambient annealing and nitridation of Ge surface (11-16) have been investigated. Variation in processing conditions and formation of an appropriate Ge/high-k gate stack has been the trend to form the transistors with Ge as channel material. For example, when a thin Al_2O_3 layer is inserted between the high-k later and Ge substrate, the equivalent oxide thickness, gate leakage current density, hysteresis, and interface state density behavior have improved significantly (17). Higher-k dielectric materials such as $Hf_{1-x}Zr_xO_2$ and/or ZrO_2 with or without Al_2O_3 are used for Ge FinFETs and nanowire transistors. With advanced ALD processing $Hf_{1-x}Zr_xO_2$ with different Hf to Zr compositions has been explored as an alternative of HfO_2 since the tetragonal phase which has higher dielectric constant can be stabilized (18). Fig. 5 shows the C-V (Fig. 5(a) and I-V (sweep) (Fig. 5(b)) characteristics of dielectric stack on Ge channel materials when the dielectric stack was exposed to slot-plane-antenna plasma oxidation (SPAO) after the dielectric deposition. It is clear from the accumulation capacitance (Fig. 5(a)) that the EOT decreased for 75% of Zr. The observation regarding the shift in location of leakage current minima (Jg_{min}) in the I-V characteristics (Fig. 5(b)) when the direction of sweep was reversed, was initially attributed to the charging and discharging of fast traps near the dielectric/Si interface. However, it is possible that the behavior could be due to ferroelectric properties of $Hf_{1-x}Zr_xO_2/Al_2O_3$ gate stack (19, 20).

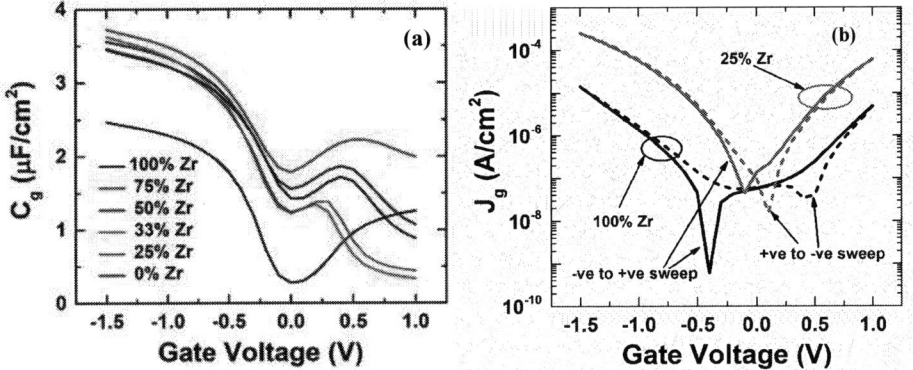

Figure 5. (a) High frequency C-V characteristics of $Hf_{1-x}Zr_xO_2/Al_2O_3$ dielectric on Ge with Zr percentage ranged from 0% to 100%; (b) gate stack leakage current behavior with change in sweep direction.

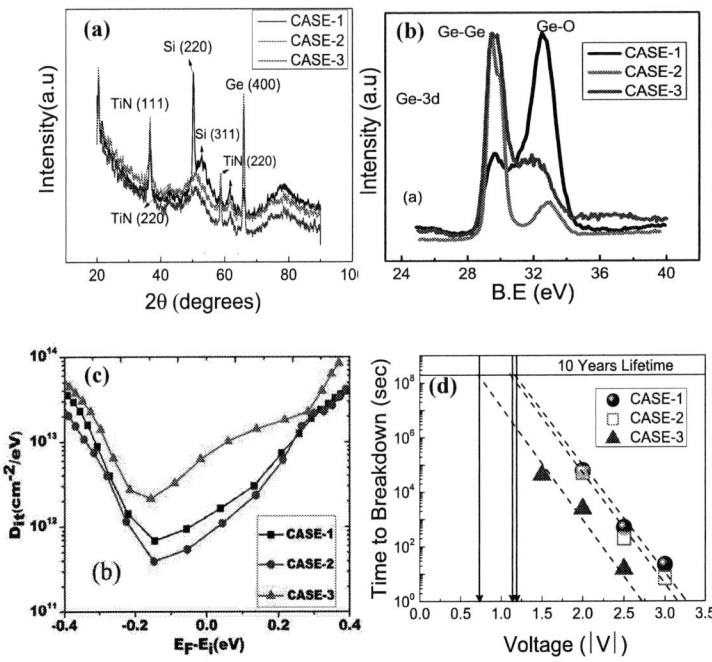

Figure 6. (a) XRD spectra, (b) High resolution XPS spectra of Ge-3d, (c) Energy distribution of D_{it} as function of bandgap, and (d) Time to breakdown as a function of different stress voltages, for all three cases.

Forming a stable GeO_x like interfacial layer by employing different dielectric processing techniques seems to improve the interface quality. Using a slot plane antenna plasma oxidation (SPAO) treatment in between the Al_2O_3 and ZrO_2 deposition process on Ge resulted in formation of a stable GeO_x layer leading to a significant improvement of the interface (21). Fig. 6 shows the dielectric characteristics and interface properties at three different process conditions: when SPAO was performed after the Al_2O_3/ZrO_2 deposition (CASE-1), In between the Al_2O_3 and ZrO_2 (CASE-2) and prior to any dielectric deposition (CASE-3). In all three cases the dielectric stack remained amorphous as shown by the XRD spectra in Fig. 6(a). The XPS spectra, shown in Fig. 6(b) suggest that the formation of a stable GeO_x interfacial layer for both CASE-1 and CASE-2. The mid-gap interface state density, D_{it}, distribution is the lowest when SPAO was performed in between the Al_2O_3 and ZrO_2 as shown in Fig. 6(c). Fig. 6(d) shows the variation of time to breakdown with the stress voltage for all the three cases indicating that SPAO in-between and after gives a superior ten-year lifetime. Furthermore, the temperature-dependent carrier transport mechanisms reveal that SPAO in-between and after dielectric deposition can effectively remove traps in high-k dielectrics and subsequently reduce the leakage current (22).

Memory Devices

Dielectric discussions for only RRAM devices are presented here. The structure of RRAM device consists of a dielectric layer sandwiched between a top and a bottom electrode. The resistance of RRAM can be reversibly switched in between high and low resistance states by the application of an external voltage because of formation of a conducting filament (low resistance state) and when the filament ruptures it returns to high resistance state. Therefore, the dielectric layer plays an important role for the RRAM switching. HfO_2 is extensively studied for the RRAM application. However, stoichiometric HfO_2 is not suitable because of lack of enough oxygen vacancies and requires high forming voltage that leads to higher power consumption (23). The multi-level storage characteristics of $HfZrO_2$ is investigated for better switching performance. In addition, top electrode material plays an important role on the switching characteristics.

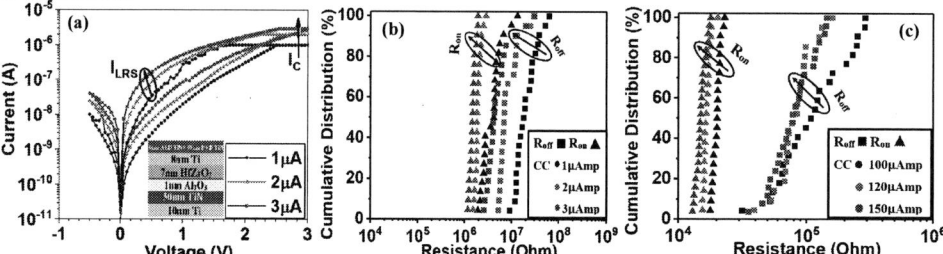

Figure 7. (a) I-V characteristics of HRS/LRS resistance $Ti/TiN/Al_2O_3/HfZrO_2/Ti/TiN$ devices for increasing compliances current; (b) The cycle-to-cycle distributions of HRS/LRS resistance of device in (a); and (c) $Ti/TiN/Al_2O_3/HfZrO_2/TiN/Ti/TiN$ devices for increasing compliances.

Fig. 7(a) shows the I-V characteristics of the three RRAM structures with the $HfZrO_2$ based RRAM as a function of increasing compliance currents. The device has a buffer layer of Al_2O_3 close to the bottom electrode and the Ti as the top electrode. Devices show increase in the low resistance state (LRS) with current levels (I_{LRS}) when the compliance current (CC) was increased. HfO_2 based devices showed poor switching characteristics (not shown), possibly due to lack of oxygen vacancies available in HfO_2 (24). When HfO_2 was alloyed with Zr, an improvement in the formation of thicker conducting filament was observed (decrease in resistance) due to increase in the number of defects sites or oxygen vacancies. Fig. 7(b) shows the cumulative distribution of cycle-to-cycle distributions on the LRS/HRS resistances of the

$HfZrO_2$ based RRAM at different compliance currents. When the top electrode was changed to TiN by depositing 2nm-thin TiN layer prior to Ti layer deposition the multi-level resistance states became distinct as shown in Fig. 7(c). But the devices with TiN required additional power as the switching compliance current increased from 1 µA to 100 µA (Fig. 7(b-c). The possible reason for the observed electrode material dependence of resistive switching behavior is due to the work function difference of the electrode material (25, 26).

Summary

In summary, we have reviewed the current trend of dielectrics in some of today's electronic devices. First, we reviewed the impact on processing conditions on dielectric quality and reliability by incorporating Zr and Al in Hf matrix. When Zr concentration was varied the reliability was improved for higher Zr concentration. With low Al concentration the reliability was improved with increasing Al concentration as it kept the Hf-matrix amorphous. Second, we have evaluated the $Hf_{1-x}Zr_xO_2/Al_2O_3$ on Ge substrate and it was observed that 75% Zr decreased the EOT and some cases $Hf_{1-x}Zr_xO_2$ showed ferroelectric-like behavior. With SPAO in-between two dielectric layers in Ge/high-k gate stack produced an excellent quality Ge-High-k interface and provided a superior ten-year lifetime of the dielectric. The multilevel switching behavior of RRAM devices with $HfZrO_2$ was studied. It was observed that intrinsic concentration of oxygen vacancies dominates the switching characteristics. Multilevel switching was found to be more prominent in case of $HfZrO_2$ based RRAM devices. Use of TiN instead of Ti as the top electrode materials enhances the multilevel storage capabilities due to the work function difference leading to oxidation at the interface.

Acknowledgment

The author would like to thank M.N. Bhuyian, Y.M. Ding and B. Jain from NJIT; G.S Kolla and N. Bhat from IISc, Bangalore, India; and K. Tapily, R. D. Clark, S. Consiglio, C. S. Wajda, and G. J. Leusink of TEL Technology Center, America, Albany, NY, for their contributions to this work.

References

1. M.T. Bohr, R.S. Chau, T. Ghani and K. Mistry, IEEE Spectrum, **44**(10), 29 (2007).
2. D.G. Schlom, S. Guha, and S. Datta, MRS Bull., **33**, 1017 (2008).
3. J. Robertson, J. Appl. Phys., **104**, 124111 (2008).
4. D.H. Triyoso, R.I. Hegde, J.K Schaeffer, D. Roan, P.J Tobin, S.B. Samavedam, B.E. White, R. Gregory, and X.-D. Wang, Appl. Phys. Lett., **88**, 222901 (2006).
5. K. Tapily, S. Consiglio, R.D. Clark, R. Vasić, E. Bersch, J. J. Sweet, I. Wells, G.J. Leusink, and A.C. Diebold, ECS Trans., **45**(3), 411- 420 (2012).
6. M.N. Bhuyian, D. Misra, K. Tapily, R. Clark, S. Consiglio, C. Wajda, G. Nakamura, and G. Leusink, ECS Journal of Solid-State Science and Technology, 3(5), N83 (2014).
7. K. Tapily, S. Consiglio, R.D. Clark, R. Vasić, C.S. Wajda, J. J. Sweet, G.J. Leusink, and A.C. Diebold, ECS Trans., **64**(9), 123 (2014).
8. Y.M. Ding and D. Misra, Journal of Vacuum Science & Technology B, **33**, 021203 (2015).
9. M.N. Bhuyian and D. Misra, IEEE Trans. on Device and Materials Reliability, **15**(2), 229 (2015).
10. M. N. Bhuyian, R. Sengupta, P. Vurikiti, and D. Misra, Appl. Phys. Lett. **108**, 183501 (2016).
11. O. J. Gregory, E. E. Crisman, L. Pruitt, D. J. Hymes, and J. J. Rosenberg, MRS Online Proc. Libr. Arch., **76**, Jan. (1986).
12. M. D. Jack and J. Y. M. Lee, J. Electron. Mater., **10**(3), 571–589 (1981).
13. Y. Wang, Y. Z. Hu, and E. A. Irene, J. Vac. Sci. Technol. A, **12**(4), 1309–1314 (1994).

14. Z. Mei, L. Renrong, W. Jing, and X. Jun, J. Semicond., **34**(6), 66005 (2013).
15. A. Dimoulas, G. Mavrou, G. Vellianitis, E. Evangelou, N. Boukos, M. Houssa, and M. Caymax, Appl. Phys. Lett., **86**(3), 32908 (2005).
16. R. Garg, D. Misra, S. Guha, IEEE Trans. of Device and Materials Reliability, **6**(3), 455-460 (2006).
17. X.-F. Li, Y.-Q. Cao, A.-D. Li, H. Li and D. Wu, ECS Solid State Letters **1**(2), N10-N12 (2012).
18. M. N. Bhuyian, P. Shao, A. Sengupta, Y. Ding, D. Misra, K. Tapily, R. D. Clark, S. Consiglio, C. S. Wajda, and G. J. Leusink, ECS Journal of Solid-State Science and Technology, **7**(2), N1-N6 (2018).
19. S. Won L.C.-M. Kim, J-H. Choi, C-M. Hyun, J-H. Ahn, Materials Letters, **252**, 56-59 (2019).
20. H. Iwai, A. Toriumi, and D. Misra, ECS Interface, **26**(4), 77-81 (2017).
21. L.G. Kolla, T.M. Ding, D. Misra and N. Bhat, J. of Vac. Sci. & Technol. B, **36**(2), 021201 (2018).
22. Y. Ding, D. Misra, K. Tapily, R. D. Clark, S. Consiglio, C. S. Wajda, and G. J. Leusink, IEEE Trans. on Device and Materials Reliability, **17**(2), 349-354 (2017).
23. D.C. Gilmer, G. Bersuker, H.-Y. Park, C. Park, B. Butcher, W. Wang, P.D. Kirsch, and R. Jammy, 2011 3rd IEEE International Memory Workshop (IMW), (2011).
24. B. Jain, C-S. Huang, D. Misra, K. Tapily, R.D. Clark, St. Consiglio, C.S. Wajda and G.J. Leusink, ECS Transactions, **89**(3), 39-44 (2019).
25. H. Shima, M. Takahashi, Y. Naitoh, and H. Akinaga, IEEE Journal of the Electron Devices Society 1 (2018).
26. G.W. Dietz, W. Antphler, M. Klee, and R. Waser, Journal of Applied Physics **78**, 6113 (1995).

ECS Transactions, 97 (5) 143-149 (2020)
10.1149/09705.0143ecst ©The Electrochemical Society

Gate Electrode Material Effect on Characteristics of Zirconium-Doped Hafnium Oxide High-*k* MOS Capacitors

W. S. Lin[1], L. Liu[1,2], Y. Kuo[1,*]

[1] Thin Film Nano & Microelectronics Research Laboratory, Texas A&M University, College Station, TX 77843, USA
[2] Electronic Science and Technology Department, Xi'an Jiaotong University, China
* yuekuo@tamu.edu

Electrical properties of MOS capacitors made of Zr-doped hafnium oxide high-*k* gate dielectric with molybdenum and copper gate electrodes have been studied. Under the thermal annealing condition, the reaction between the gate electrode and gate dielectric is highly dependent on the electrode material. The *C-V* characteristics of the molybdenum gated capacitor are improved with the post metal annealing due to the reduction of defects. However, the copper gated capacitor deteriorated after the annealing step due to the diffusion of copper into the gate dielectric layer. Therefore, the metal gate material is critical to the electrical properties of the MOS capacitor with the high-*k* gate dielectric film.

Introduction

The high-*k* dielectric has been used to replace SiO_2 as the gate dielectric material in the scaling down of the metal-oxide-semiconductor (MOS) device. While the leakage current of the MOSFET with an ultrathin thermal SiO_2 gate dielectric, e.g., < 1.0 nm, has a very large leakage current density (*J*) of greater than $1 A/cm^2$, the device made of a proper high-*k* gate dielectric of the same equivalent oxide thickness (EOT) can have a leakage current density several orders of magnitude lower (1, 2). This is due to the high-*k* film's large band gap energy and large band offset with Si, which generates a potential barrier over 1 eV (3-5). However, one of the reliability concerns for the Hf-based dielectric is the low crystallization temperature, e.g., < 600°C (6). It was reported that the Zr-doped HfO_2 (ZrHfO) thin film has many advantages over the undoped HfO_2 thin film, such as higher crystallization temperature, higher permittivity, lower interface thickness, and lower interface density of states (7-9). Previously, polycrystalline Si (poly-Si) was used as the gate electrode in MOSFETs because it can stand over 800°C, and is compatible with SiO_2. However, in the case of the high-*k* gate dielectric material, it has poor bonding with the poly-Si gate electrode and forms an interface with high defects, which trap charges and reduces free charge carriers (10). By replacing poly-Si with a metal gate electrode, a high free carrier density, e.g., $1 \times 10^{20}/cm^3$, can be achieved and the phonon scattering effect can be decreased, which reduces the mobility degradation (10).

MOS capacitors with the sputter deposited ultra-thin ZrHfO gate dielectric layer, i.e. EOT < 1 nm, and a metal gate electrode have been successfully demonstrated previously (2). The capacitor's electrical properties, such as charge trapping capacity, interface state density, leakage current, and breakdown voltage, can be improved with the addition of other elements, such as molybdenum (Mo) or ruthenium (Ru) (11). The choice of the metal gate material can affect the threshold voltage of the device because of the work function effect (12). Mo has been proved to be a promising alternative gate electrode material to SiO2 in the formation of a high quality interface (13). On the other hand, in addition to the high conductivity and high electromigration resistivity (14), copper (Cu) has a similar work function as that of Mo (15), which makes it a possible gate electrode material. In this paper, two kinds of metal gates, i.e., Cu and Mo, are separately deposited on top of the ZrHfO gate dielectric film to study the MOS capacitor properties.

Experimental

MOS capacitors were fabricated on dilute hydrofluoric acid cleaned p-type (10^{15} cm^{-3}) Si (100) wafers. The ZrHfO film was sputter deposited from a Zr/Hf (12/88 wt.%) target in Ar/O2 (1:1) at 10 mTorr and 80 W for 12 min. The sample was subsequently treated with a post deposition annealing (PDA) step at 800°C under N2 atmosphere for 5 min. Then, a 200 nm thick Cu or Mo electrode film was sputter deposited on top of the ZrHfO film at 10 mTorr and 80 W, i.e., 20 min for the former and 120 min for the latter. They were etched into round gate electrodes of 200 μm, 300 μm and 400 μm diameters defined with a lithography step. The Mo gate was wet etched with the (H3PO4/HNO3/CH3COOH/H2O 80:5:5:10 vol.) solution. The Cu gate capacitor was etched with a room-temperature plasma-based process that is composed of the HCl plasma chlorination of the Cu film followed by the HCl solution dipping (16). After the photoresist was stripped, the sample was treated with a post metal annealing (PMA) step at 250°C under N2 atmosphere for 10 min. Samples without the PMA treatment, i.e., controlled samples, were also prepared and characterized for comparison.

The capacitance-voltage (C-V) curve of the capacitor was measured with the Agilent 4284 LCR meter. The EOT was calculated from the equation of

$$EOT = \frac{\varepsilon_0 \varepsilon_i A}{C_{ox}}$$ [1]

where ε_i is the dielectric constant of SiO2 (~3.9), ε_0 is the permittivity of the free space (8.85×10^{-12} F/m), A is the area of the gate electrode, and C_{ox} is the oxide capacitance extracted from the C-V curves in the accumulation region. Electrical parameters, such as the flat band voltage (V_{FB}), interface density of states (D_{it}), and oxide trapping density (Q_{ot}) were extracted from the C-V curve using the NCSU CVC program (17). The Q_{ot} was calculated by the following equation (18):

$$Q_{ot} = \frac{(C_{acc} \times \Delta V_{FB})}{q}$$ [2]

where q is the electronic charge (1.6×10^{-19} C), C_{acc} is the accumulation capacitance/area, ΔV_{FB} is the magnitude of the flat band voltage shift.

The elemental composition and chemical states of the gate electrode/high-k interface was characterized with the X-ray photoelectron spectroscopy (XPS) after stripping the gate electrode from the MOS capacitor.

Results and Discussion

Mo gated MOS capacitors

Figure 1 shows the C-V hysteresis curves of Mo gated capacitors (a) without and (b) with the PMA step. The samples were measured with the gate voltage (V_g) swept from -3 V to 3 V (forward) and then back to -3 V (backward). All curves show the counterclockwise hysteresis phenomenon. The sample without the PMA treatment shows a large flat-band voltage shift (ΔV_{FB}) of 1.4881 V while the annealed sample shows a much smaller ΔV_{FB} of 0.083 V. The V_{FB} of the forward C-V curve of the sample without the PMA was -2.052 V. It changed to -0.056 V when the device experienced the PMA treatment. Therefore, the low temperature PMA step reduced the hole trapping capability of the dielectric layer. In addition, the EOTs of the samples with and without PMA treatment are 11.2 nm and 20.9 nm, respectively. The Q_{ot}, which was calculated from the metal-semiconductor work function and flat band voltage of the C-V curve, decreased with the application of the PMA treatment from 9.61×10^{11} to 1.09×10^{11} cm^{-2}. The PMA step lowered the charge trapping density of the capacitor. The large oxide trapping density of the capacitor was probably from the long sputter deposition time of the Mo gate electrode. According to ref. 19, the increase of the sputtering voltage or sputtering time decreased the E_{00} value that is an important tunneling parameter in the metal-silicon junction since the kT/E_{00} value represents the relative importance of thermionic emission and tunneling (19). It indicates the increase of the number of charge trapping centers at the interface. The surface of the high-k layer can be damaged by hot electrons during the long Mo sputtering time, i.e. 120 min (20). Therefore, a large number of hole-like defects were generated in the high-k film or at the high-k/Si interface. In addition, it was reported that the Mo gate could react with high-k film to form new bonds that reduced the work function (21). The barrier height between the gate electrode and the gate dielectric is critically dependent on the work function or the Fermi level of the gate metal. It affects the tunneling effect between the gate electrode and oxide traps in high-k layer (22). The trapped electrons are more likely to stay in place and compensate the positive trapped charges when the level lies below the Fermi level of the metal. As a result, the 250˚C annealing step is effective in reducing the charge trapping density.

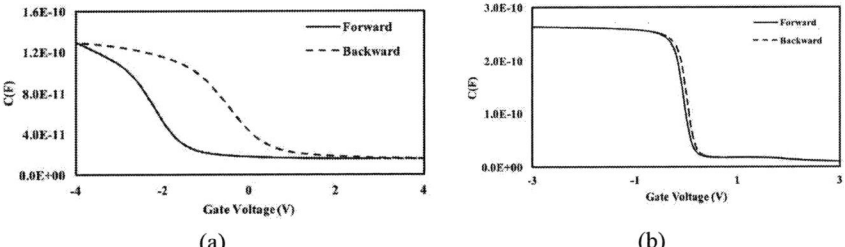

(a)   (b)

Figure. 1. C-V hysteresis curves of 400 μm diameter Mo gated capacitors (a) without and (b) with the PMA treatment. Measured at 1 MHz.

The PMA treatment also affects the leakage current of the device. Figure 2 shows J-V curves of Mo gated capacitors (a) without and (b) with the PMA step. For each device, the 1st measurement caused the hard breakdown of the dielectric film, which formed nano-resistors (23, 24). The 2nd measurement curve confirmed the permanent formation of these conductive paths, i.e., the leakage current was much larger than that of the 1st measurement curve. In the 1st curve, as a negative gate voltage (V_g) is applied, the holes were injected from the p-Si substrate to the dielectric layer to generate defects. The leakage current was very low at the beginning. When the magnitude of the V_g was increased, the leakage current increased until reaching the voltage (V_{BD}) where it jumped by several orders of magnitude abruptly. The dielectric film was broken due to the connection of defects (11, 25). The magnitude of V_{BD} of the device without the PMA was larger than that with the PMA, i.e., -13.6 V vs. -5.65 V. Also, the leakage current of the former was lower than that of the latter. These results are consistent with the EOT numbers.

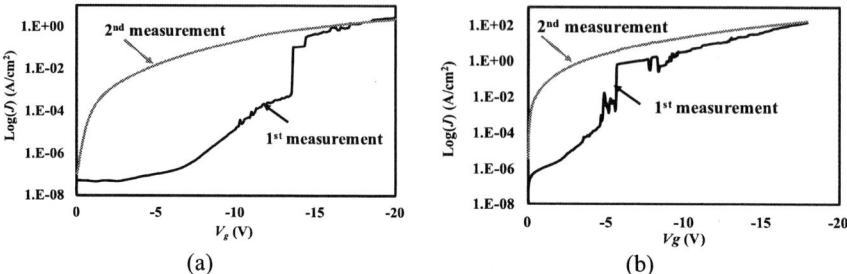

(a)   (b)

Figure 2. *J-V* curves of 400 μm diameter Mo gated capacitors (a) without and (b) with the PMA treatment.

Cu gated MOS capacitors

Figure 3 shows *C-V* hysteresis curves of Cu gated capacitors (a) without and (b) with the PMA step. Before PMA, as shown in Fig. 3(a), the *C-V* curve was similar to that of a good capacitor with V_{FB} of -0.357 V (forward) and -0.373 V (backward), EOT of 9.3 nm, and Q_{ot} of 1.07×10^{10} (cm^{-2}). The Q_{ot} is related to the presence of unsaturated bonds, vacancies, and unsatisfied atomic valences in the dielectric layer (26). During the device operation, charge carriers, i.e., electrons and holes, may be trapped and attached to defects in the gate dielectric film, which further affects the gate-induced electric field at the dielectric/Si interface and raises the threshold voltage. However, the *C-V* curve of the PMA treated sample, i.e., in Fig. 3(b), showed the distorted shape. It resembled the low-frequency measured curve, i.e., formation of inversion layer at the $+V_g$. Also, the capacitance in the accumulation region was much lower than that of the device without the PMA treatment. Therefore, the PMA step affected the capacitor's material properties.

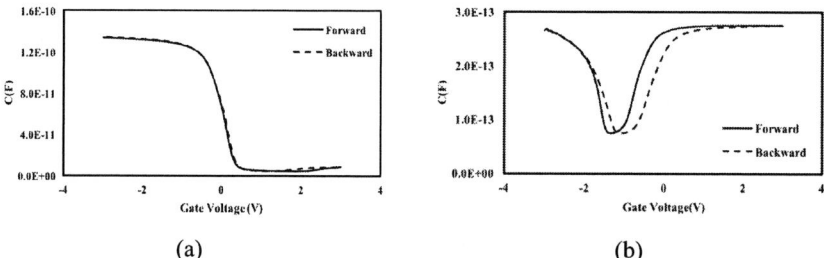

(a)   (b)

Figure 3. *C-V* hysteresis curves of 400 μm diameter Cu gated capacitors (a) without and (b) with the PMA treatment. Measured at 1 MHz.

Since Cu can easily diffuse through a dielectric film (14), it is possible that Cu diffused to the high-k gate dielectric during the PMA step. Figure 4 shows the chemical states of Cu at the Cu/ZrHfO interfaces of capacitors (a) without and (b) with the PMA treatment. The peak at 933 eV corresponds to the binding energy of Cu $2p^{3/2}$ (933 eV) in the elemental and/or oxidized states (27). It was reported that under thermal stress between 150°C and 300°C, copper atoms could penetrate through the thermal oxide into the bulk silicon to cause catastrophic effect on the capacitor characteristics (14). They induce electric-defect centers near the interface and deteriorate the C-V characteristics. They can also generate deep-level states at the silicon/silicon dioxide interface, which results in the increase of the generation-recombination rate of minority carriers and the reduction of the minority carrier lifetime(14). Therefore, the capacitance change of the Cu gated capacitor is related to the existence of Cu atoms in the high-k dielectric layer.

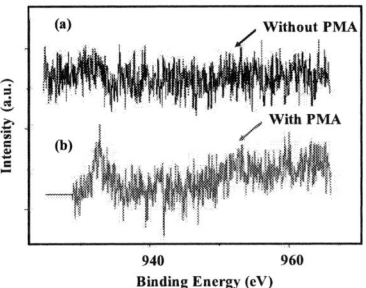

Figure 4. XPS spectra of 400 μm diameter Cu gated devices (a) without and (b) with PMA treatment.

J-V curves of Cu gated capacitors were also affected by the diffusion of Cu to the high-k film. Figure 5 shows J-V curves of Cu gate capacitors (a) without and (b) with the PMA treatment. The 1st measurement in Fig. 5(a) shows several breakdown steps due to many defects in the gate dielectric film. The first breakdown occurred at around -6 V. However, the 1st measurement in Fig. 5(b) shows one clear breakdown step at around -4.1 V. Some original defects in the gate dielectric film were probably removed by the PMA treatment. On the other hand, the leakage current of the PMA treated capacitor was larger than that of without the PMA treatment. This is consistent with the XPS measurement that Cu atoms were diffused into the gate dielectric layer to cause high leakage current and to facilitate the breakdown process.

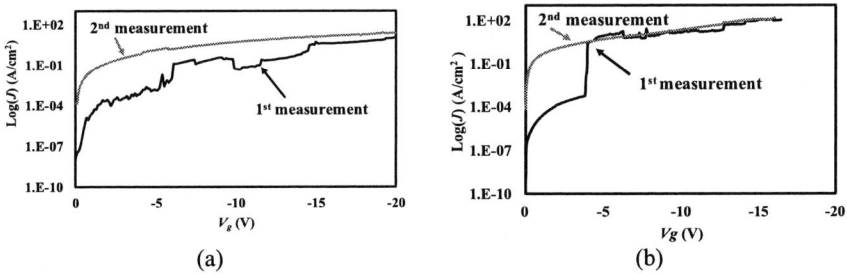

Figure 5. Current density-voltage (J-V) of 400 μm diameter Cu gated capacitors (a) without and (b) with PMA treatment.

Conclusion

Electrical properties, i.e., *J-V* and *C-V* characteristics, on 400 μm diameter Mo and Cu gated capacitors fabricated with the thermal annealing procedure were studied. For dielectric parameters, the properties of high-*k* layer in Mo gated devices were improved by PMA process due to the reduction of defects in the high-*k* film while those of Cu gated devices degraded from metal diffusion in the dielectric layer. As shown in *J-V* curves, the breakdown voltages of both metal gated devices decreased upon the application of PMA. The performance of the capacitor is influenced by the PMA treatment, which affects the reaction between the gate and the high-*k* gate dielectric as well as the removal of defects in the film and at the interface. Therefore, the selection of gate material is crucial to fabricate proper MOS capacitor.

Acknowledgments

Lingguang Liu acknowledges the financial support of National Natural Science Foundation of China (Grant Number 61771382) and the Shaanxi International Science and Technology Cooperation and Exchange program (2018KW-034).

References

1. C.-H. Lin and Y. Kuo, *Electrochem. & Solid-State Lett.*, **13**, H83 (2010).
2. J. Yan, Y. Kuo and J. Lu, *Electrochem. & Solid-State Lett.*, **10**, H199 (2007).
3. S.-H. Lo, D. Buchanan, Y. Taur and W. Wang, *IEEE Electron Device Lett.*, **18**, 209 (1997).
4. J. Robertson, *J. Vac. Sci. Technol. B*, **18**, 1785 (2000).
5. S. Mohsenifar and M. Shahrokhabadi, *Microelectronics and Solid State Electronics*, **4**, 12 (2015).
6. B. H. Lee, L. Kang, R. Nieh, W.-J. Qi and J. C. Lee, *Appl. Phys. Lett.*, **76**, 1926 (2000).
7. Y. Kuo, *ECS Trans.*, **3**, 253 (2006).
8. Y. Kuo, J. Lu, J. Yan, T. Yuan, H. C. Kim, J. Peterson, M. Gardner, S. Chatterjee and W. Luo, *ECS Trans.*, **1**, 447 (2006).
9. W. Luo, T. Yuan, Y. Kuo, J. Lu, J. Yan and W. Kuo, *Appl. Phys. Lett.*, **89**, 072901 (2006).
10. A. Verma, A. Mishra, A. Jha and K. Verma, *Intl. J. Adva. Rese. Electrical Electron.*, **3**, 9667 (2014).
11. C.-C. Lin and Y. Kuo, *J. Vac. Sci. Technol. B*, **31**, 030605 (2013).
12. K. H. Jing, M. M. Arshad, A. Huda, A. Ruslinda, S. C. Gopinath, R. Ayub, M. Fathil, N. Othman and U. Hashim, in *AIP Conference Proceedings*, p. 020073 (2016).
13. Q. Lu, R. Lin, P. Ranade, Y. C. Yeo, X. Meng, H. Takeuchi, T.-J. King, C. Hu, H. Luan and S. Lee, in *IEDM*, p. 641 (2000).
14. Y. Shacham - Diamand, A. Dedhia, D. Hoffstetter and W. Oldham, *J. Electrochem. Soc.*, **140**, 2427 (1993).
15. T. J. Drummond, *J. Appl. Phys.* (1999).
16. Y. Kuo and S. Lee, *Appl. Phys. Lett.*, **78**, 1002 (2001).

17. D. G. Seiler, A. C. Diebold, W. M. Bullis, T. J. Shaffner, R. Mcdonald and E. J. Walters, *Characterization and Metrology for ULSI Technology*, American Institute of Physics, New York, United States (1998).
18. M. Houssa, M. Naili, C. Zhao, H. Bender, M. Heyns and A. Stesmans, *Semicond. Sci. Technol.*, **16**, 31 (2001).
19. B. L. Sharma, *Metal-Semiconductor Schottky Barrier Junctions and Their Applications*, p. 42, Springer (1984).
20. S. Berg, L. Andersson, H. Norström and E. Grusell, *Vac.*, **27**, 189 (1977).
21. D. Ha, P. Ranade, Y.-K. Choi, J.-S. Lee, T.-J. King and C. Hu, *Jpn. J. Appl. Phys.*, **42**, 1979 (2003).
22. V. Afanas' ev and A. Stesmans, *J. Appl. Phys.*, **95**, 2518 (2004).
23. Y. Kuo and C.-C. Lin, *Appl. Phys. Lett.*, **102**, 031117 (2013).
24. Y. Kuo, *ECS Trans.*, **69**, 23 (2015).
25. Y. Kuo and C.-C. Lin, *Appl. Phys. Lett.*, **102**, 031117 (2013).
26. D. Fleetwood, P. Winokur, R. Reber Jr, T. Meisenheimer, J. Schwank, M. Shaneyfelt and L. Riewe, *J.Appl. Phys.*, **73**, 5058 (1993).
27. M. C. Biesinger, *Surf. Interface Anal.*, **49**, 1325 (2017).

Graphitic Nanoporous Carbon Thin Films: Fabrication Method, Structural, Electrical and Gas Sensor Properties

O. M. Slobodian[a], Yu. V. Gomeniuk[a], A. V. Vasin[a,b], A. V. Rusavsky[a,b], P. N. Okholin[a], O. Yo. Gudymenko[a], O. Yu. Khyzhun[c], A. S. Nikolenko[a], P. M. Lytvyn[a], A. Korchovyy[a], R. Yatskiv[d], T. M. Nazarova[b], V. G. Stepanov[a], D. V. Kisyl[a], and A. N. Nazarov[a,b]

[a] Lashkaryov Institute of Semiconductor Physics NAS of Ukraine, Kyiv, Ukraine
[b] National Technical University "Igor Sikorsky KPI", Kyiv, Ukraine.
[c] Frantsevych Institute for Problems of Materials Science NAS of Ukraine, Kyiv, Ukraine
[d] Institute of Photonics and Electronics CAS, Praha, Czech Republic
correspondent author email: nazarov@lab15.kiev.ua

New method (magnetron plasma enhanced CVD) is suggested to obtain record values of porosity of thin carbon films. Further high temperature annealing results in graphitization of the carbon film and reduction of their resistivity down to 10^2 Ohm×cm. Plasma treatment in forming gas leads to modification both of a surface and bulk thin nanoporous graphitic film that decreases resistivity of the film, reduces sensor sensitivity to water vapour and enhances sensor sensitivity to ammonia gas.

Introduction

Nanoporous carbon film is a very attractive material for different fields of applications such as catalysis, energy storage, purification of gases and liquids and chemical gas sensors (1). For many applications the porous carbon film has to be conductive, for example, in biomedical electronic devices and chemical sensors operating on electrical resistance response. Combination of porosity and conductivity in such films is a great advantage but it is a big challenge as well. High conductivity in carbon is associated with high concentration of sp^2 bonds and highly percolated graphitic/graphenic phases. To obtain graphitic phase in the porous carbon films a high temperature annealing in vacuum or nitrogen atmosphere can be used (2). In the present work we propose an original method of "magnetron plasma enhanced chemical vapour deposition" of low density amorphous carbon. It will be shown that poor conductivity and chemical sensitivity of as-deposited films can be strongly enhanced by high temperature thermal annealing. Additionally, low-temperature RF plasma treatment (3) in forming gas ambient is proposed to increase resistive sensitivity of the material to ammonia.

Experimental

Porous Carbon Films Preparation

Deposition of low-density amorphous carbon films was performed using specific deposition regime of typical planar RF-magnetron with silicon target. At first deposition stage a very thin "adhesion" layer (few nanometers) was deposited on the substrate using pure argon and discharge power 200 W. After "adhesion" layer is formed the deposition process was modified by reducing of discharge power down to 40-120 W, and introducing of either methane or acetylene in the chamber. At such parameters no silicon sputtering occurred on the silicon target and the deposition process is determined primarily by plasma enhanced decomposition of methane/acetylene in the near-substrate region. Above described deposition mode is something similar to "low power/pressure" PECVD process and was named as magnetron PECVD. Si(100) wafers and Si wafers with 200 nm silicon oxide layer grown by thermal oxidation were used as substrates.

Deposited a-C:H films were modified by thermal annealing ("graphitization") for enhancement of electrical conductivity. "Graphitization" has been performed with thermal annealing in N_2 atmosphere at temperatures up to 650 °C for 5min. Samples #2, fabricated from methane, after graphitization were subjected by low-temperature RF plasma treatment at power density of 0.5 W/cm^2 for 30 seconds in different gas ambient: forming gas (10% H_2+90% N_2); 100% N_2; 100% Ar. Four kinds of the carbon film, deposited using different discharge power and methane or acetylene flow rate were studied (see Table I).

TABLE I. Sample Labels, Corresponding Deposition Conditions and Structural Parameters of the Carbon Films.

Samples	Working Gas, Ar/Reactive Gas (sccm)	Power (W)	Reactive Gas	Density of Carbon Film Before/After Annealing, (g/cm^3)	Porosity of Carbon Film After Annealing (%)
#1	14/6.0	70	CH_4	1.34/1.48	35
#2	15/4.1	122	CH_4	1.30/1.58	30
#3	15/2.0	41	C_2H_2	1.33/1.08	52
#4	15/8.0	41	C_2H_2	1.36/1.01	55

Methods of Characterization

Chemical composition of as-deposited films, high temperature annealed films and low-temperature RF plasma treated ones was examined by X-ray photoelectron spectroscopy (XPS) using the UHV-Analysis-System (SPECS Surface Nano Analysis Company). The XPS spectra were recorded using excitation by an X-ray Mg Kα source (E=1253.6eV). A flood gun was used in the present XPS experiments to overcome charging effects.

The X-ray reflectometric (XRR) technique was used to measure a density and thickness of the carbon films, and performed with PANalytical X'Pert the Pro MRD XL diffractometer using CuKα1 radiation (0.15405 nm). Surface morphology was studied by atomic force microscopy (AFM) with NanoScope IIIa Dimension 3000 (USA).

Four point probe method was used for the primary analysis of electrical properties. More careful study was performed by conductance vs. frequency measurements. The gas sensitivity was examined by measurement of resistance between Ni contacts. Measurements were performed with semiconductor parameter analyzer Agilent 4156C and Agilent E4284A Precision LCR meter.

Results and Discussion

Chemical Bonds in the Carbon Films

The survey XPS spectra are presented in Fig.1. A significant amount of oxygen is detected in both as-deposited and thermal annealing samples (see Fig. 1(a) and Table II).

However, the maximum of the O1s core-level spectra in all samples is located at about 532.5eV that corresponds rather to oxygen-containing atmospheric species adsorbed on the carbon films. Furthermore, as can be seen from data listed in Table II, the thermal annealing of the carbon film causes decreasing the content of oxygen on their surfaces by about 30%. It is worth noting that Si is not present in the XPS spectra. The high resolution XPS C 1s core-level spectra of the as-deposited carbon films (not shown here) reveal the C 1s spectra with their maxima at about 285.0eV, which is close to the C 1s binding energy of diamond. Annealing at 650°C results in the C 1s spectra shift to about 284.4eV that well corresponds to that of the C 1s spectrum in graphite (4). Results of deconvolutions of C 1s spectra for as-deposited and post annealed films, presented in Table II, allow for suggesting that as-deposited carbon films are characterized by C–C bonds with dominant sp^3 hybridization, while annealing leads to transformation of four-fold coordination into three-fold coordination with dominant sp^2 hybridization.

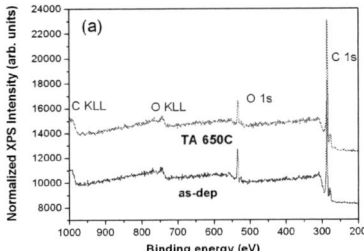 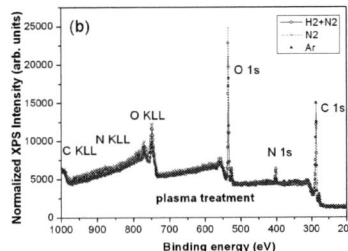

Figure 1. XPS data obtained for the porous carbon films (sample #2): (a) – survey spectra of the as-deposited carbon film and after thermal annealing in N_2 atmosphere at 650°C; (b) – survey spectra of the thermal annealed carbon film after plasma treatment at specific power of 0.5W/cm^2 for 30 sec.

TABLE II. Composition (at %) of the Carbon Film Surfaces for Sample #2 as Determined by XPS measurements .

Carbon film	Carbon (%) (sp^2/sp^3)	Oxygen (%)	Nitrogen (%) (N-sp^2C/N-sp^3C)
As-deposited	68.6 (40.9/43.4)	31.4	-
Thermal annealed (650°C)	78.7 (77.0/6.2)	21.3	-
Plasma treated by H_2+N_2	54.3	39.2	6.5 (65.3/30.1)
Plasma treated by N_2	51.1	43.0	5.9 (53.0/34.2)
Plasma treated by Ar	68.2	30.1	1.7 (64.4/29.6)

The survey XPS spectra for plasma treated samples are presented in Fig. 1(b). Formation of the nitrogen bonds and increasing of oxygen bonds concentration are observed. The high resolution XPS N 1s core-level spectra of the plasma treated carbon films (not shown here) reveal the N 1s spectra with their maxima at about 400.3 eV, which is close to the (sp^2)C-N bond. Results of deconvolution of the spectra are presented in Table II which demonstrates inclusion of a considerable amount of nitrogen into lattice of graphite structure. The increase of oxygen concentration in the carbon film after plasma treatment is due to residual oxygen in vacuum camera.

Structure and Surface Morphology of the Carbon Films

The film density and porosity were determined using XRR technique. The film density was determined using critical angle, when X-ray wave can penetrate into film. Thickness of the thin film on the substrate can be determined using interference pattern of two reflected beams (from the film surface and film–substrate interface). Using thickness values and Snell's law the densities of the carbon films before and after annealing were estimated (2). From data presented in Table 1 we can conclude that as-deposited carbon films have low density (from 1.30 to 1.36 g/cm^3) which for the carbon films, deposited in methane atmosphere (sample #1 and #2), is increased a little after high-temperature annealing (up to 1.58 g/cm^3). However, in case of carbon films deposited in acetylene atmosphere (samples #3 and #4) the density of the films becomes considerably lower (up to 1.01 g/cm^3) after the thermal annealing. Comparing the density of our carbon films with the value for tetrahedral amorphous carbon (ta-C: 3.26 g/cm^3 (5)) and graphite (2.26 g/cm^3 (6)) we can estimate porosity of the film $\Phi(\%)$ using simple equation:

$$\Phi(\%) = [1 - (\rho_m / \rho_c)] \times 100\%, \qquad [1]$$

where ρ_m and ρ_c are the densities of measured and reference materials, respectively. Porosity of as-deposited films was calculated using the density of ta-C as reference, while for annealed film the density of graphite was used. Results, presented in Table I, show that porosity of as-deposited films is about 59% and reduced down to 30% after annealing of carbon film deposited in methane but it remains large enough. In case of carbon films deposited in acetylene atmosphere the thermal annealing results in a slight decrease of the film porosity which can reach about 55% for graphitic film that is the record value for such kind of material (7).

Surface morphology of the graphitic films fabricated from methane and treated by plasma was studied by AFM method. Topography maps of the carbon films before and after plasma treatment for sample #2 are illustrated in Fig.2. Surface morphology of the carbon film is characterized by RSM roughness of about 1.65 nm and exhibits nanoscale granular topography. Surface of the films contains a large amount of holes/pores with size of few tens of nanometers (black regions in inset of Fig.2 (a, b)). After plasma treatment the surface is smoothened and RMS roughness decreases down to 1.25 nm for the sample treated in argon plasma. The reduction in half-width of peak of histogram of surface heights (Fig. 2 (c)) quantitatively illustrates the smoothing of the surface relief after plasma treatment. Pure nitrogen plasma smoothes less than pure argon and forming gas plasma.

Electrical Properties of the Carbon Films

Resistivity data for the samples #1, #2, #3 and #4, measured by four point probe method, before and after annealing at 350, 550 and 650°C are presented in Fig. 3(a). Resistivity of the films synthesized in methane and annealed at 650°C (samples #1 and #2 in Fig. 3(a)) decreases by 5 orders of magnitude as compared to as-deposited samples, and reaches the value up to 1×10^4 Ohm×cm. In case of carbon films synthesized in acetylene (samples #3 and #4 in Fig. 3(a)) their resistivity after the thermal annealing is reduced to 1×10^2 Ohm×cm. Thus graphite films synthesized in acetylene have record values of a porosity and low resistivity.

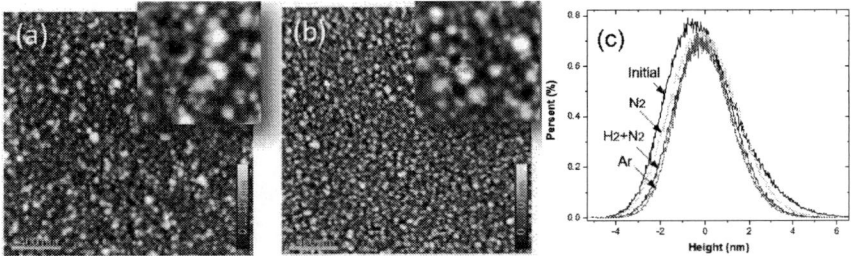

Figure 2. Deconvolution of AFM images for annealed at 650°C (a) and forming gas plasma treated (b) carbon films (Sample #2). (c) Histogram of surface heights for initial and plasma treated samples (for "zero" is taken as the average surface level).

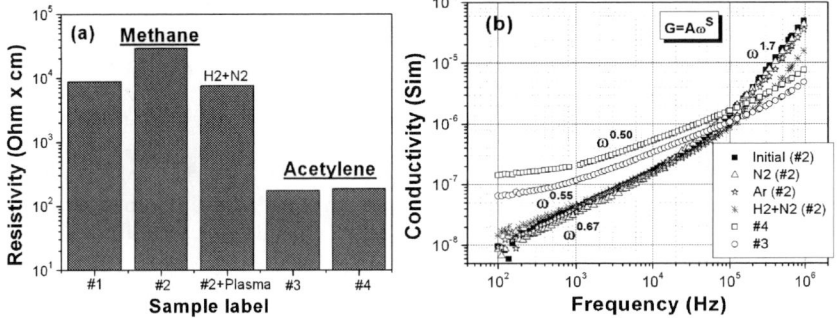

Figure 3. (a) Resistivity of the carbon films after high-temperature annealing and additional plasma treatment in forming gas; (b) dependences of conductivity vs. frequency for samples #2, #3 and #4.

Dependencies of electrical conductivity on frequency demonstrate a power law ($G=A\omega^S$) for graphite films synthesized both in methane and acetylene (see Fig. 3(b)) but with different power exponent. The S equals 0.67 in a frequency range from 1×10^2 to 1×10^4 Hz for sample #2, synthesized in methane, and 0.50 for samples #3 and #4, synthesized in acetylene that corresponds to hopping mechanisms of electron transport in the films. It should be noted that plasma treatment in Ar or N gas atmosphere does not change the conductivity of the carbon films, but a plasma treatment in forming gas ($10\%H_2+90\%N_2$) resulted in increasing of the conductivity at low frequency and reduction of S to 0.55 (Fig. 3(b)). The first effect can be associated with hydrogen passivation of electricaly active defects inside the film, and the second one - with changing of the phonon frequency of the carbon structure.

Sensor properties

Response on chemical ambient was examined by measurement of the electric current between a pair of contacts at fixed signal level (1V). Results for such analytes as ammonia (NH_3) and water vapour are presented in Fig.4 for samples synthesized in methane and subjected to plasma treatment. The carbon films after high temperature annealing show enough high resistance sensitivity to ammonia (about 20 %) and to water vapour (about 12 %). However after plasma treatment (especially in N_2 and forming gas atmosphere) the sensitivity to water vapour is reduced down to 5 %. Resistance sensitivity to ammonia increases on first step up to 30%, and the adsorption rate is also noticeably enhanced.

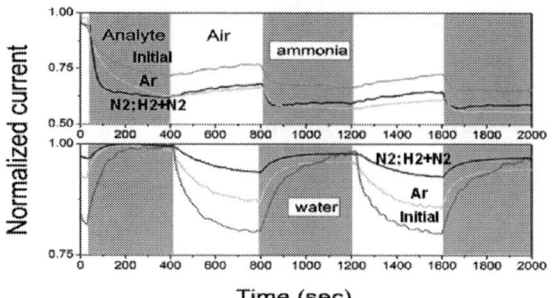

Figure 4. Time dependences of normalized current (for 1V applied voltage) for #2 sample, measured under (a) ammonia and (b) water vapour before and after plasma treatments.

Conclusions

Record value of porosity (55%) and small resistivity (100 Ohm×cm) of graphitic film synthesized in ambient acetylene were obtained. It was shown that plasma treatment in forming gas allows us not only to modify the surface of the films but also changes its conductivity at low frequencies. Plasma treatment in forming gas and nitrogen results in reduction of sensitivity to water vapour and increases sensitivity and response rate to ammonia gas.

Acknowledgments

The work was supported by Ministry of Education and Science of Ukraine (Project 2211-F).

References

1. N. A. Travlou1, M. Seredych, E. Rodriguez-Castellon and T. J. Bandosz, *J. Mater. Chem. A*, **3**, 3821-3831 (2015).
2. O.M. Slobodian, A.V. Rusavsky, A.V. Vasin, O.Yu. Khyzhun at all. *Applied Surface Science*, **496**, 143735 (2019).
3. A.N. Nazarov, V.S. Lysenko and T.M. Nazarova, *Semiconductor Physics, Quantum Electronics & Optoelectronics,* **11**(2), 101 (2008).
4. O. Khyzhun, E. Zhurakovsky, A. Sinelnichenko and V. Kolyagin, *Journal of Electron Spectroscopy and Related Phenomena,* **82,** 179 (1996).
5. A. Libassi, A.C. Ferrari, V. Stolojan, B.K. Tanner et al., *MRS Proceedings*, **593** (1999).
6. J.C. Rivière and S. Myhra, *Handbook of surface and interface analysis: methods for problem-solving,* 2nd ed., CRC Press, Boca Raton (2009).
7. S. Kim, B. B. Sahu, B. M. Weon, J.G. Han et al., *Jap. J. Appl. Phys.,* **54**, 010304 (2015).

CHAPTER 4

Memory and Circuits

Influence of Current Redistribution in Switching Models
for Perpendicular STT-MRAM

S. Fiorentini[1], R.L. de Orio[1,2], S. Selberherr[2], J. Ender[1], W. Goes[3], and V. Sverdlov[1]

[1]Christian Doppler Laboratory for NOVOMEMLOG at the
[2]Institute for Microelectronics, TU Wien, Gußhausstraße 27-20, 1040 Vienna, Austria
[3]Silvaco Europe Ltd., Cambridge, United Kingdom
e-mail: fiorentini@iue.tuwien.ac.at; phone: +43 15880136001

> Simulation of switching in spin-transfer torque magnetoresistive random access memory is usually performed by assuming that the torque is created by a position- and time-independent current density. However, in real circuits the voltage is fixed, not the current density. The assumption of a fixed current density, especially in modern devices with a tunneling magnetoresistance up to 200%, becomes thus questionable. In this work we compare the switching time distribution obtained under the assumptions of fixed voltage and fixed current density for a wide range of tunneling magnetoresistance and surface area values. We demonstrate that the approximate fixed current density approach can reproduce the correct switching times, provided that the current value is appropriately adjusted. We show that the correction on the current depends on the switching speed, dictated by different system parameters.

Introduction

The outstanding improvement in the performance of modern integrated circuits is supported by the continuous down-scaling of semiconductor devices. However, this leads to a substantial increase in leakages, which results in growing stand-by power consumption. A viable path to mitigate these issues is the introduction of non-volatility in integrated circuits. Spin-transfer torque (STT) magnetoresistive random access memory (MRAM) is a promising candidate (1-6). It is competitive with conventional non-volatile flash memory as it combines high speed, excellent endurance, and low costs. The range of potential STT-MRAM utilization goes from automotive and Internet-of-Things applications to embedded memories and last level caches (7). The core of an STT-MRAM device is a magnetic tunnel junction (MTJ), where the relative orientation of its magnetic layers provides a way of storing binary information. Switching between the two possible configurations is achieved by passing an electric current through the structure (8,9). The electrons become spin-polarized by the reference layer (RL) and, when entering the free layer (FL), act via the exchange interaction on its magnetization by exerting a torque. With a sufficiently large current density, the magnetization of the free layer can be flipped. The usual approach for micromagnetic simulations of STT switching is to assume a constant and uniform current density (10). In circuits and applications, however, the voltage, rather than the current density, is fixed during the switching process. As the tunneling resistance in an MTJ depends on the changing relative magnetization orientation of the two magnetic layers, the current depends on time. Moreover, the magnetization of the free layer is non-uniform

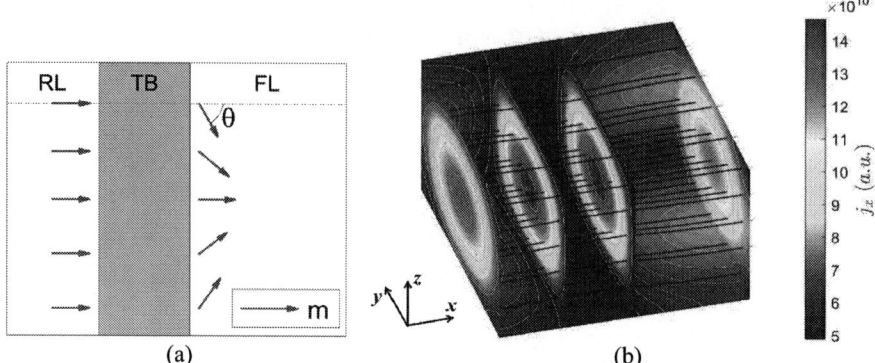

(a)   (b)

Figure 1. (a) Schematization of an MTJ structure with non-uniform free layer (FL) magnetization. TB is the tunnel barrier and RL the reference layer. (b) Non-uniform distribution of the current density in the MTJ. The current flows towards the paths of minimum resistivity, where the magnetization vectors in the two layers are aligned.

during switching, which results in a position- (and time-) dependent current density $\mathbf{J}(\mathbf{r}, t)$ (Fig. 1). The assumption of a constant and uniform current density is thus violated, especially in devices with tunneling magnetoresistance ratios (TMR) of about 200% and higher (11). In order to clarify, if the fixed current density assumption for switching time evaluation can be still used, we compare its results with switching at a fixed voltage. We also consider a description in which the total current is fixed, but the current density is locally determined by the magnetization alignment and the corresponding TMR value.

STT-MRAM Model

In modern MRAM cells the binary information is stored as the relative orientation of the magnetic layers in an MTJ, which consists of a sandwich of two ferromagnetic layers and an insulating tunneling layer. The magnetization in the free layer (FL) can switch, while the magnetization in the second, reference layer (RL) is fixed by the exchange coupling to a pinned layer (12). CoFeB is typically used for the magnetic layers, while MgO is the typical material for the insulating layer, as it provides a good TMR. The TMR is defined as

$$TMR = \frac{G_P - G_{AP}}{G_{AP}}, \qquad [1]$$

where G_P and G_{AP} are the conductances in the parallel and anti-parallel states, respectively. Another beneficial property of this choice of materials is the interface coupling between MgO and CoFeB, which renders the ferromagnetic layers perpendicularly magnetized. In this configuration, the thermal relaxation path and the switching path coincide, leading to lower switching currents as compared to structures with in-plane magnetization. The development of tools which are able to properly simulate the switching process and the torques acting on the magnetization can improve the design of novel MRAM devices. The magnetization dynamics is described by the Landau-Lifshitz-Gilbert (LLG) equation. With a spin transfer torque term added, the LLG equation for the free layer reads as (13)

$$\frac{\partial \mathbf{m}}{\partial t} = -\gamma\mu_0 \mathbf{m} \times \mathbf{H}_{\text{eff}} + \alpha \mathbf{m} \times \frac{\partial \mathbf{m}}{\partial t} + \frac{1}{M_S}\mathbf{T_S} \qquad [2a]$$

$$\mathbf{T_S} = \gamma\frac{\hbar}{2e}\frac{0.5\,J_C\,P}{d\left(1+P^2\cos\theta\right)}\mathbf{m} \times (\mathbf{m} \times \mathbf{x}), \qquad [2b]$$

where γ is the gyromagnetic ratio, μ_0 is the vacuum permeability, $\mathbf{m}=\mathbf{M}/M_S$ is the position-dependent normalized magnetization in the free layer, M_S is the saturation magnetization in the free layer, α is the Gilbert damping factor, $\hbar$ is the reduced Plank constant, e is the electron charge, J_C is the magnitude of the current density, P is the spin current polarizing factor (13), which is assumed equal in the two magnetic layers for this work, d is the thickness of the free layer, θ is the angle between local magnetization vectors in the free and reference layers, $\mathbf{x}$ is the direction of the magnetization in the reference layer, and $\mathbf{H}_{\text{eff}}$ is the effective magnetic field, containing different contributions, namely the external field, the exchange interaction, the anisotropy field, the Ampere field, the demagnetizing field, and the stray field from the reference layer. In order to simulate the switching at a finite temperature, a stochastic thermal contribution to $\mathbf{H}_{\text{eff}}$ is also included.

The usual approach for STT-switching simulations is to assume a constant and uniform value for J_C. In order to test this assumption in an MTJ with TMR=200% and non-uniform magnetization in the free layer, we compute the current density flowing through the structure as

$$\mathbf{J}_C = -\sigma\nabla V, \qquad [3]$$

where V is the electric potential and σ is the conductivity. The potential in the ferromagnetic leads is computed by solving the Laplace equation $\nabla^2 V = 0$. The local conductance of the barrier, taken as suggested in (14) as

$$G(\theta) = \frac{G_P + G_{AP}}{2}\left(1+\left(\frac{TMR}{2+TMR}\right)\cos\theta\right), \qquad [4]$$

is imposed as a boundary condition on the ferromagnet/insulator interface. In Fig. 1b the results for the configuration schematized in Fig. 1a are shown. As a result of the non-uniform conductance of the structure, the current density is highly non-uniform too. Thus, it is necessary to evaluate the impact of assuming a fixed voltage on the simulation of magnetization reversal.

Results

We compare a realistic approach in which the *voltage* during switching is kept constant with the *fixed current density* approach. In addition, the fixed voltage and fixed current density models are compared to an approach (15), generalized to p-MTJs, in which the *total current* is fixed, but redistributed according to the position-dependent resistance determined by the local relative magnetization orientation in the two magnetic layers. Herein, the value of the current in the fixed current/current density models is equal to the voltage in the fixed voltage model divided by the resistance in the *initially* parallel (P) or anti-parallel (AP) state. The free layer is perpendicularly magnetized. The switching time depends on the realization of the stochastic magnetic field, which mimics the magnetization fluctuations at room temperature. It is demonstrated that, by slightly increasing the current

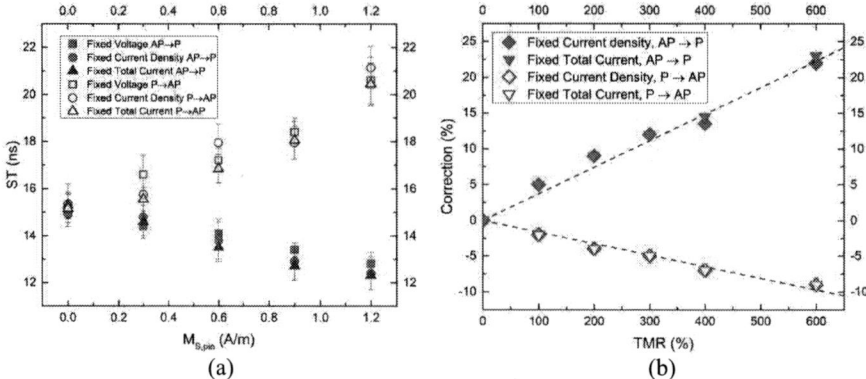

(a)   (b)

Figure 2. (a) Switching times (ST) for the three approaches as a function of the stray field from the reference layer, modeled by its saturation magnetization (M_S), after applying the current correction. Error bars represent the thermal distribution. (b) Dependence of the current correction on the TMR for both P→AP and AP→P switching, for T=300 K. The dashed lines represent a linear fit.

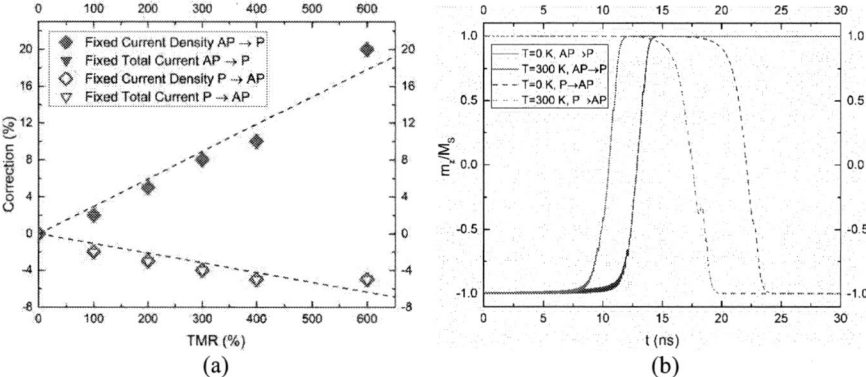

(a)   (b)

Figure 3. (a) Dependence of the current correction on the TMR for both P→AP and AP →P switching, for T=0 K. (b) Comparison between switching realizations for the fixed voltage approach at T=0 K and T=300K.

from its initial value for AP to P switching and decreasing it for P to AP switching, the switching time distributions can be matched (16) for any value of the stray field induced by the reference layer (Fig. 2a). Here, we systematically study the dependence of this current correction on system parameters. The dependence of the current correction on the TMR at room temperature is reported in Fig. 2b. It is observed that the value of the correction increases with higher TMR. We then performed simulations at zero temperature, and in this case, the required correction to reproduce the fixed voltage results is lower than the one at room temperature, while it still increases with the TMR (Fig. 3a). However, the switching time at zero temperature is also longer than at room temperature, as shown in Fig. 3b. This provides a strong indication that the correction to the current is not universal and depends on the system parameters. To elaborate on the physical origin of this dependence we performed macrospin simulations with the free layer represented by a single cell. The initial magnetization direction is slightly tilted from its perfect perpendicular orientation in order to reduce the incubation time of switching. By gradually increasing the tilting angle, one can monitor the dependence of the current correction on the switching time, as reported in Fig. 4a. The data show that a faster switching requires a higher correction to the current value, which explains the difference between the room and zero temperature simulations. As the switching is faster at room temperature, the correction required on the current is also higher. The macrospin results can also help to explain the origin of the current correction. Fig. 4b reports switching realizations for the fixed voltage and the fixed current density models with increasing values of the correction. The

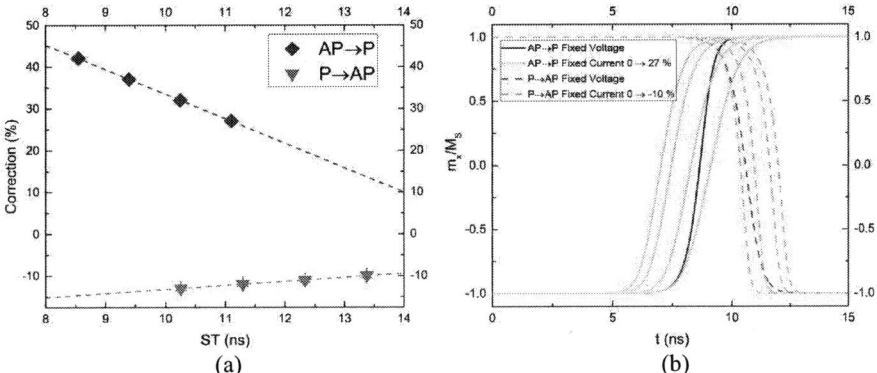

(a)   (b)

Figure 4. Macrospin simulations to explain the origin and behavior of the current correction. In (a) the dependence of the current correction on the switching time is reported. Shorter switching times require a larger current correction. In (b) we show how the correction affects the switching realization, for both AP→P and P→AP. The different slope of the fixed current approach is compensated by the current correction.

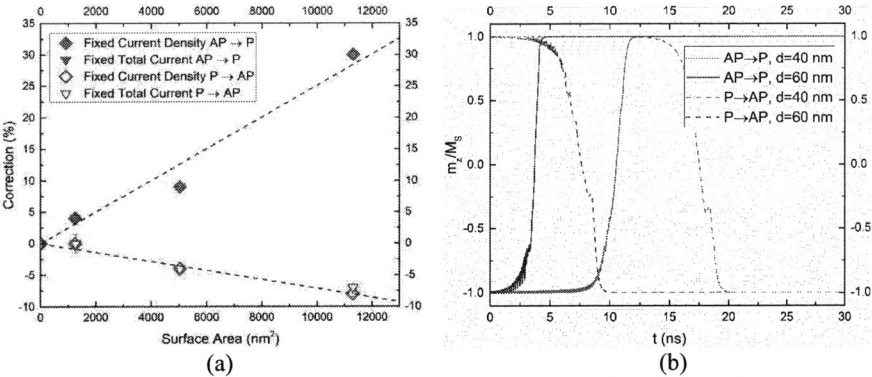

(a)   (b)

Figure 5. (a) Dependence of the current correction on the surface area of the structure at T=300 K, for both P→AP and AP→P switching. The dashed lines represent a linear fit. (b) Comparison between switching realization for the fixed voltage approach with a structure diameter of d=40 nm and d=60 nm.

correction is necessary for the switching process in the fixed current density approach to begin earlier for AP→P and later for P→AP, in order to compensate the difference in the slope to the fixed voltage approach. As an additional test of the dependence of the correction on system parameters, we performed simulations for different values of the surface area of the structure at room temperature. The results are reported in Fig. 5. The required current correction increases with the surface area (Fig. 5a). This implies that the switching in a structure with a larger diameter is faster, and this is indeed observed in Fig. 5b in agreement with the macrospin results.

Conclusion

We performed simulations of switching in an STT-MRAM structure. We compared the switching time distribution obtained under the assumption of a fixed voltage during switching to the ones obtained under the approximation of fixed current density and fixed total current across the structure. We showed that it is possible to reproduce the switching times obtained within the realistic fixed voltage approach with the approximate fixed current density approach for a wide range of system parameters, provided that the dependence of the current correction on such parameters is taken into consideration.

Acknowledgments

This work was supported by the Austrian Federal Ministry for Digital and Economic Affairs and the National Foundation for Research, Technology and Development.

References

1. S.Aggarwal, H.Almasi, M.DeHerrera *et al.*, *Proceedings of the IEDM*, 2.1.1 (2019).
2. K.Lee, J.H.Bak, Y.J.Kim *et al.*, *Proceedings of the IEDM*, 2.2.1 (2019).
3. V.B.Naik, K.Lee, K.Yamane *et al.*, *Proceedings of the IEDM*, 2.3.1 (2019).
4. J.G.Alzate, U.Arslan, P.Bai *et al.*, *Proceedings of the IEDM*, 2.4.1 (2019).
5. G.Hu, J.J.Nowak, M.G.Gottwald *et al.*, *Proceedings of the IEDM*, 2.6.1 (2019).
6. W.J.Gallagher, E.Chien, T.W.Chiang *et al.*, *Proceedings of the IEDM*, 2.7.1 (2019).
7. S.Sakhare, M.Perumkunnil, T.H.Bao *et al.*, *Proceedings of the IEDM*, 420 (2018).
8. J.C.Slonczewski, *Journal of Magnetization and Magnetic Materials* **159**, L1 (1996).
9. L.Berger, *Physical Review B* **54**, 9353 (1996).
10. A.Makarov, T.Windbacher, V.Sverdlov, and S.Selberherr, *Semiconductor Science and Technology* **31**, 113006 (2016).
11. W.Skowronski, M.Czapkiewicz, S.Zietek, *et al.*, Scientific Reports **7**, 10172 (2017).
12. S.Bhatti, R.Sbiaa, A.Hirohata, *et al.*, *Materials Today* **20**, 530 (2017).
13. A.Makarov, *Ph.D. Thesis*, Institute for Microelectronics, TU Wien, Vienna (2014).
14. J.Slonczewski, *Physical Review B* **71**, 024411 (2005).
15. D.Aurelio, L.Torres, and G.Finocchio, *Journal of Magnetization and Magnetic Materials* **321**, 3913 (2009).
16. S.Fiorentini, R.Orio, W.Goes *et al.*, *Proceedings of the SISPAD*, 57 (2019).

Readout Circuit Design Using Experimental Data of Line-TFET Devices

W. Gonçalez Filho[a], R. S. Rangel[b], J.A. Martino[a] and P.G.D. Agopian[a,c]

[a] LSI/PSI/USP, University of Sao Paulo, Sao Paulo, Brazil
[b] ECE, University of Toronto, Toronto, Canada
[c] UNESP, Sao Paulo State University, Sao Joao da Boa Vista, Brazil
Email: walter.filho@usp.br

By considering the analog characteristics of Line Tunneling Field Effect Transistors (Line-TFETs) that are suitable for small-signal amplification, this paper studies the design of a readout circuit with these devices while making comparisons with conventional MOSFET designs. The results show that the Line-TFET design exhibits high gain and low reading error (51dB open loop gain) while using a simple one-stage amplifier and results in a huge reduction in circuit area by using pseudo feedback resistors that have their differential resistance increased for smaller dimensions, achieving up to 50Gohm in a 120nm x 100nm device. This enables cutoff frequencies below 1Hz while using nanometer devices and smaller capacitors. Moreover, the readout circuit achieves 33nW of power consumption even though the Line-TFET devices are not biased in the subthreshold regime.

Introduction

Different applications in modern analog circuit design, ranging from Internet of Things (IoT) to biosignal acquisition, require integrated circuits with minimum power consumption and circuit area. The line-tunneling transistor (Line-TFET) (1), an emerging device fabricated at imec/Belgium, presents characteristics that could help to overcome some of the issues that arise with the very tight design constraints imposed by modern analog applications. This device presents extremely high intrinsic voltage gain, which simplifies circuit topology, and low current levels that are suitable for low-power and low-frequency applications (2)(3). Moreover, TFETs have been shown to be resistant to noise (4) and temperature variation (5).

Readout circuits are frequently necessary in order to realize signal amplification of the outputs of a sensor or a sensor array within a determinate bandwidth. Circuit area is especially critical in implantable sensors, for instance, or in applications that use large sensor arrays, as the readout circuit must be replicated for each node (6). The reading error is inversely proportional to the amplifier's open loop gain, which is an important constraint when performing the acquisition of small signals (7). This fact results in a compromise between reading error and circuit area, as short MOSFET devices present small intrinsic voltage gain. Furthermore, in some applications of biosignal acquisition, such as EEG or ECG, the bandwidth of the bandpass amplifier must cover very small frequencies, thus often requiring big capacitance and resistance values in the closed-loop configuration, which greatly increases circuit area (6)(7). Therefore, this work proposes the use of Line-TFET devices in order to perform the design of a readout circuit with high gain (above 45dB, open-loop configuration) and bandwidth from f<1Hz (common mode suppression) to about f=100Hz.

Device Description

The Line-TFET fabricated at imec/Belgium is depicted in Figure 1. The $Si_{0.55}Ge_{0.45}$ source is heavily doped with boron and extends along the entire gate length. Between the source and the gate, there is a very thin (about 5nm thick), ideally undoped strained silicon region, the pocket. In this device, tunneling happens from the source valence band to the pocket conduction band evenly along the horizontal direction (hence line-tunneling). Therefore, tunneling generation rate and, consequently, the drain current, are proportional to the gate area in this device.

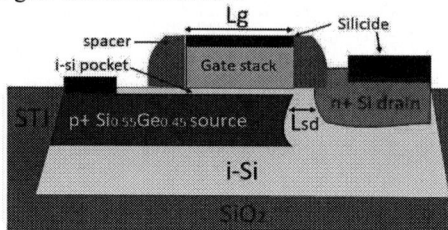

Figure 1 – Line-TFET device structure.

Figure 2 depicts the Ids x Vgs transfer characteristics of the Line-TFET (Lg=1μm) and of a 130nm conventional MOSFET (Lg=130nm). These gate lengths maximize the current level for both devices. Figure 2 shows that the Line-TFET achieves lower current levels than the MOSFET device. Furthermore, the Line-TFET requires higher supply voltage because of its delayed ON set voltage, degraded subthreshold slope due to trap and phonon assisted tunneling (1) and parasitic source-to-drain tunneling that degrades the saturation characteristics for low Vgs (2). However, the Line-TFET presents the possibility of greatly reducing circuit area for a given power budget thanks to its much higher intrinsic voltage gain that simplifies circuit topology, lower current levels for smaller gate dimensions, which is suitable for low-frequency design, and reduction of feedback resistors and capacitors in the closed-loop configuration.

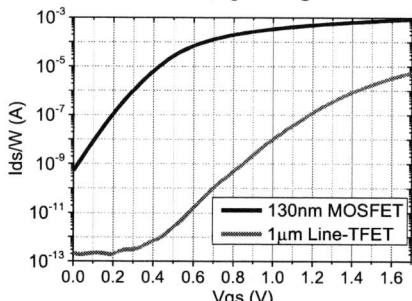

Figure 2 – Drain current as a function of the gate voltage of the Line-TFET (Lg=1μm) and of a conventional MOSFET, 130nm process (Lg=100nm).

Line-TFET device modeling was performed using a lookup table with experimental data coded in Verilog-A for the DC characteristics while the parasitic capacitances (Cgs, Cgd and Cgg=Cgs+Cgd) were estimated through TCAD simulation and also implemented in a lookup table. In order to simplify the analysis, P and N-type

Line-TFET transistors are considered symmetric and gate leakage current was disregarded. The circuit was also designed with a 130nm conventional MOSFET technology PDK in order to make comparisons with the Line-TFET design.

Readout Circuit Design

Figure 3.a) shows the readout circuit including the pseudo-resistor structure of the feedback resistors, using gate-source connected Line-TFET devices (8) and diode-connected MOSFETs, while Figure 3.b) shows the transistor level implementation of the operational amplifier. Pseudo resistors in the Line-TFET design require two transistors in parallel because of the unidirectionality of the current in these devices (8), while in the MOSFET design 6 diode-connected devices in series were necessary to increase the feedback resistance. The gate dimensions of the input transistors are 100nm x 1μm and 250nm x 4μm for the Line-TFET and MOSFET designs, respectively.

Passive common-mode feedback was implemented by using pseudo resistors (M_{cmfb}) in the same configurations that were previously mentioned for each technology. These transistors must have as large differential resistance as possible in order to preserve the output resistance and voltage gain. For the MOSFET design, this means increasing the gate length and, consequently, circuit area. Sufficient differential resistance was achieved by using one transistor with gate dimensions of 150nm x 10μm for each output. In the Line-TFET case, smaller transistors result in greater differential resistances. Therefore, two 120nm x 100nm transistors in parallel were used for each output.

Pseudo resistors were also used for the implementation of the bias current source (M_{bias1}). The MOSFET design uses one N-type and one P-type diode-connected transistor, with gate dimensions 150nm x 4μm, while the Line-TFET design uses one diode-connected device with dimensions 200nm x 1μm. The Line-TFET implementation should provide robustness to temperature variation if the devices are biased with sufficiently high gate voltages (Vgs greater than about 1.2V) (5). In this bias region, band-to-band tunneling, which depends little on the temperature, is much more significant than TAT, which is a thermally activated transport mechanism (5). The ratio between M_{bias2} and M_5 dimensions determines the proportion between the bias current in each branch of the operational amplifier circuit. In the Line-TFET design, this proportion is equal to (W_{bias2} x L_{bias2}) / (W_5 x L_5). Therefore, we used (230nm x 100nm) / (100nm x 1μm) in order to reduce the current in the bias branch. This was not possible in the MOSFET design because $M_{1,2}$ already had very low drain currents by using 250nm x 4μm devices.

a)   b)

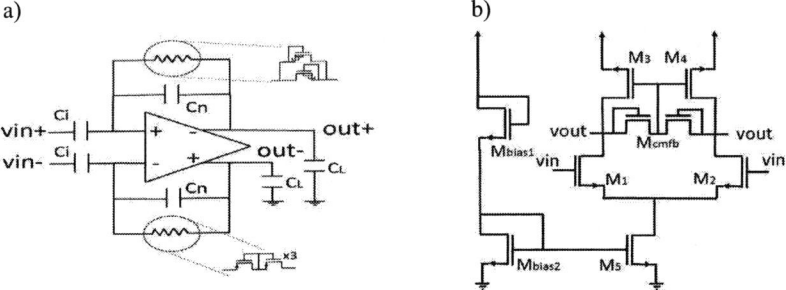

Figure 3 – a) Readout circuit and pseudo resistors implementation b) Transistor implementation of the amplifier.

Figure 4 depicts the Bode diagram of gain and phase comparing both designs, while Table I summarizes the results. In the MOSFET design, the output and bias connections draw larger current, thus consuming more power. The lower transistor efficiency and higher Vdd for Line-TFETs hinder its power consumption and the required supply voltage. The Line-TFET is prone to degradation of the saturation-like region for low Vgs, especially for small devices, due to source-to-drain parasitic tunneling, and to transistor efficiency degradation because of trap and phonon assisted tunneling (2). However, the Line-TFET design presents 51dB open-loop DC gain, reducing the reading error by a factor of 2 in comparison with the MOSFET design (45dB open-loop gain), while using a simple topology. The ~1Hz to 100Hz bandwidth was achieved with both devices, however the Line-TFET design required smaller transistors due to its lower current level and the proportionality of the drain current with the gate area. Circuit area was greatly reduced when designing pseudo-resistors with Line-TFET devices as their drain current is proportional to the gate area, increasing the differential resistance of small devices. Therefore, it achieves up to 50 Gohm in a 120nm x 100nm device, while the MOSFET design required 6 150nm x 10µm transistors in series for the implementation of each feedback resistor. This characteristic of the Line-TFET enables cut of frequencies below 1Hz while using nanometer devices and smaller capacitances (Ci, Cn and Cl).

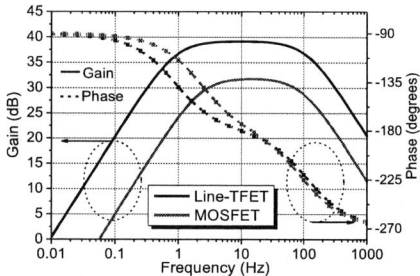

Figure 4 – Closed loop Bode diagram of gain and phase for the Line-TFET and MOSFET designs.

TABLE I. Results comparing Line-TFET and MOSFET designs.

	Line-TFET	MOSFET
Vdd	2.5V	1.2V
Power	33nW	730nW
Closed loop gain	39dB	31dB
Relative reading error	Av_{MOS}/Av_{Line}=0.4	
Bandwidth	850mHz~117Hz	2.15Hz~105Hz
Ci/Cn/C_L	10pF/150fF/1pF	100pF/1.5pF/100pF
W_{bias1} x L_{bias1}	200nm x 1µm	150nm x 4µm (x2)
W_{bias2} x L_{bias2}	230nm x 100nm	150nm x 4µm
$W_{1,2}$ x $L_{1,2}$	100nm x 1µm	250nm x 4µm
$W_{3,4}$ x $L_{3,4}$	100nm x 1µm	1.25µm x 4µm
W_{CMFB} x L_{CMFB}	120nm x 100nm (x4)	150nm x 10µm (x2)
$W_{feedback}$ x $L_{feedback}$	120nm x 100nm (x2)	150nm x 10µm (x6)
Total resistor area	0.096µm²	21µm²

Table II exhibits the results of this work in comparison with other readout circuit designs. Even with higher supply voltage and not biasing the devices in the subthreshold region, the Line-TFET design achieves similar power consumption of some nW thanks to the lower current level of this device, making it easier to perform very low-frequency signal amplification. Greater intrinsic voltage gain of Line-TFETs allow the simplification of the topology by using a simple one-stage amplifier with low reading error. Using smaller capacitors and pseudo-resistors instead of a dc servo loop (DSL) in order to define a mHz high-pass corner frequency was also possible because of the extremely high differential resistance of small gate-source connected Line-TFETs.

TABLE II. Comparison with other readout circuit designs.

	Line-TFET	(6) MOS 65nm	(9) MOS 65nm	(10) MOS 180nm
Vdd	2.5V	0.6V	0.6V	0.45V
Power	33nW	1nW	16.8nW	730nW
Closed loop gain	39dB	32dB	51-96dB	52dB
Bandwidth	850mHz~117Hz	1.5Hz~350Hz	<0.5Hz~250Hz	0.25Hz~10kHz
Ci/Cn/C_L	10pF/150fF/1pF	Ci=40pF Cn=1pF	Ci=50pF Cn=0.5pF	-
Amplifier topology	One-stage	Two-stage	Two-stage	Two-stage
Feedback topology	Pseudo-resistors (0.096μm²)	DSL	DSL	Pseudo-resistors

Conclusions

This work proposes the use of Line-TFET devices in the design of a low power, low-frequency readout circuit. Transistor modelling was performed with lookup tables obtained from experimental data for the DC characteristics, obtaining a model that considers several non-ideal phenomena, like source-to drain tunneling and trap and phonon assisted tunneling, in a very straightforward way. When comparing the Line-TFET transfer characteristics with a 130nm conventional MOSFET, the Line-TFET is shown to be less power efficient and to present lower drain current. However, the lower current levels of these devices facilitate very low-frequency design while reducing circuit area. Furthermore, the proportionality of the drain current with gate area in Line-TFETs also helps reducing circuit area, which is a critical figure of merit for the design of readout circuits in some particular applications.

The fact that the drain current is proportional to the gate area in Line-TFETs implies that the differential resistance of this device increases for smaller dimensions. Moreover, the TFET off state current could be used to implement very high resistances with gate-source connected devices, even though it requires two devices in parallel because of the unidirectionality of the current (8). Line-TFET devices achieve up to 50Gohm differential resistance with nanometer transistors (gate dimensions of 120nm x 100nm), greatly reducing circuit area when implementing pseudo resistors in the bias circuit, common-mode feedback and feedback loop in comparison with MOSFETs, which required up to 6 150nm x 10μm transistors in series.

As a result of the aforementioned considerations, a readout circuit with bandwidth from ~1Hz to 100Hz that was designed with Line-TFETs achieved 33nW of power consumption while reducing the reading error by a factor of two in comparison with the MOSFET design and a 220-fold reduction in pseudo resistor area. Therefore, this device may be suitable for low-power, low-frequency analog applications, such as readout circuit design for IoT applications like biosignal monitoring or applications with large sensor arrays that must minimize circuit area and power consumption.

Acknowledgements

The authors thank CAPES, CNPq and FAPESP for the financial support and imec for supplying the devices.

References

1.   A. M. Walke, A. Vandooren, R. Rooyackers, D. Leonelli, A. Hikavyy, R. Loo, A. S. Verhulst, K. Kao, C. Huyghebaert, G. Groeseneken, V. R. Rao, K. K. Bhuwalka, M. M. Heyns, N. Collaert, and A. V. Thean, *Transactions on Electron Devices*, p. 707-715, **61** (3) (2014).
2.   W. Gonçalez Filho, E. Simoen, R. Rooyackers, C. Claeys, N. Collaert, J. A. Martino, and P. G. D. Agopian, *Semiconductor Science and Technology*, accepted for publication (2020).
3.   A. Acharya, A. Solanki, and S. Dasgupta, *Transactions on Electron Devices*, p. 322-330, **65** (1) (2018).
4.   F. S. Neves, P. G. D. Agopian, J. A. Martino, B. Cretu, R. Rooyackers, A. Vandooren, E. Simoen, A. V. Thean, and C. Claeys, *Transactions on Electron Devices*, p. 1658-1665, **63** (4) (2016).
5.   M. D. V. Martino, J. A. Martino, P. G. D. Agopian, R. Rooyackers, E. Simoen, N. Collaert, and C. Claeys, *Semiconductor Science and Technology*, **33** (075012) (2018).
6.   P. Harpe, H. Gao, R. V. Dommele, E. Cantatore, and A. H. M. van Roermund,, *Journal of Solid State Circuits*, p. 240-248, **51** (1) (2016).
7.   L. Zhong, X. Lai, and D. Xu, *Journal of Solid State Circuits*, p. 2240-2251, **53** (8) (2018).
8.   A. R. Trivedi, S. Carlo, and S. Mukhopadhyay, *50th ACM/EDAC/IEEE DAC*, (2013).

Proton-Irradiation Influence on Current Mirror Circuit Using Verilog-A Based on Experimental SOI FinFET Characteristics

B. R. de Sousa [a], P. G. D. Agopian [a,b] and J. A. Martino [a]

[a] LSI/PSI/USP, University of Sao Paulo, Sao Paulo, Brazil
[b] UNESP, Sao Paulo State University, Sao Joao da Boa Vista, Brazil
correspondent author email: brunarsousa@usp.br

In this work, a simple methodology is proposed to simulate the current mirror circuit based on the triple-gate SOI FinFET experimental data, called lookup table in Verilog-A. It was analyzed the reliability of the model implemented, comparing between experimental and simulated data, which has proven to be reliable. It was also evaluated the performance of the transistor and as well the circuit regarding the efficiency and the gain, for p- and n-types, based on three different fin widths, before and after proton-irradiation.

Introduction

To be able to build systems using a semiconductor device, it is common to use simulation programs in research stage, that can predict the electrical behavior of the device by its physical conditions and bias applied (1). Since the triple-gate SOI FinFET does not have an accurate first order analytic model to be collected from physical simulation programs, it is proposed to use a method called lookup table in Verilog-A, consisting of a detailed experimental characterization of the device aiming the construction of a lookup table with the data to be used in simulations (1–4).

It is known that the SOI technology provides significant immunity to single-event phenomena when compared to bulk transistors, due to the thin active silicon region that provides a better electrostatic coupling and the existence of the buried oxide beneath the channel that isolates the transistor active area from the substrate. Still, the radiation effect have to be considered because the trapped charges in the buried oxide (BOX) degrade the transistor characteristics due to the higher Total Ionization Dose (TID) (5–9).

Devices Characteristics

The studied devices are p- and n-type triple-gate SOI FinFETs fabricated in imec, Belgium. They were processed on SOI substrates with a thick buried oxide (t_{BOX}) of 150 nm. The gate dielectric of the devices consists of 2 nm HfSiON on 1 nm SiO_2 interfacial layer, resulting in an Equivalent Oxide Thickness (EOT) of 1.5 nm. The gate is 10 nm TiN covered by 100 nm poli-Si. The fin height (h_{FIN}) is 65 nm, the channel length (L) is 150 nm and three different fin widths (W_{FIN}) were evaluated: 20 nm, 120 nm and 870 nm. Each transistor has 5 fins in parallel.

The proton irradiation has been performed at the Cyclone facility in Louvain-la-Neuve (Belgium). The beam energy is 60 MeV up to a fluence of 10^{12} p/cm^2. No bias was applied during the irradiation and all the devices are unpackaged.

Figure 1 shows a schematic view of a triple-gate SOI MOSFET (FinFET) structure.

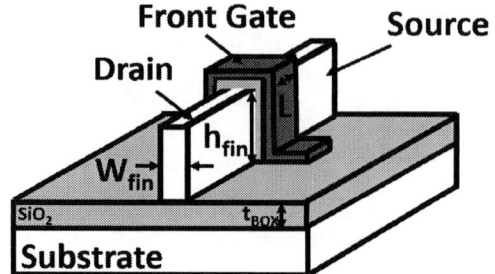

Figure 1. Triple-gate SOI MOSFET (FinFET) structure.

Results

Figure 2 shows the experimental drain current (I_{DS}) as a function of the front gate voltage (V_{GF}) for different fin widths before and after the proton irradiation. It is possible to notice that although narrow devices ($W_{FIN} = 20nm$) showed to be almost immune to proton irradiation due to a higher coupling between gates, the wider devices present a degradation in subthreshold swing (SS) for n-FinFETs while for p-FinFET the SS improves. The reason is that the radiation induces positive charges in the buried oxide, resulting in a reduction of the threshold voltage at the back interface (V_{th2}), which is good for p-FinFET and bad for n-FinFET, as explained in (8–10).

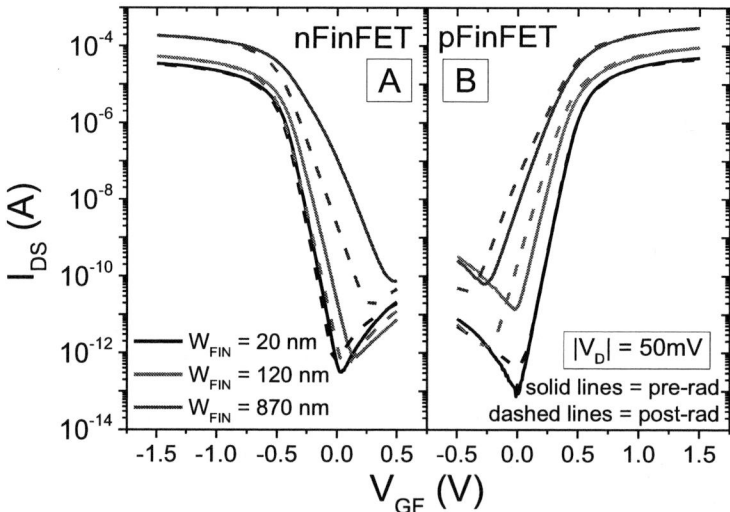

Figure 2. Experimental drain current as a function of the front gate voltage for (A) p- and (B) n-FinFETs, for different fin widths, pre and post-radiation.

Figure 3 shows the drain current (I_{DS}) as a function of the drain voltage (V_{DS}) for different fin widths before and after the irradiation, for simulated and experimental data. It is possible to verify that since the first measured data inserted in the lookup table have small steps, it is possible to simulate the devices behavior even for another bias variation (V_{DS} instead of V_{GF}) and still achieve a good agreement. Since the simulation values accurately fit the experimental data, it is reliable to use the lookup table model using Verilog-A to simulate the real device behavior using this method.

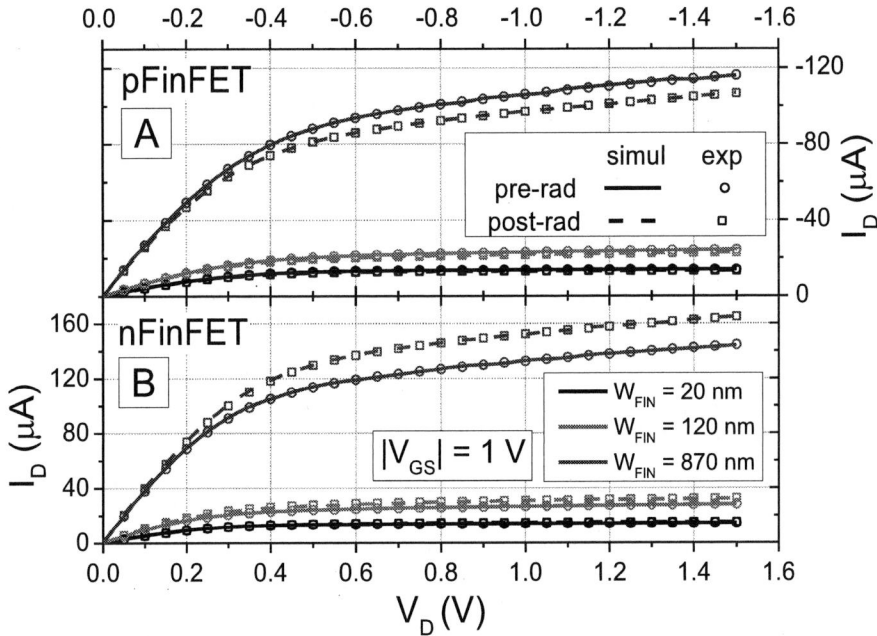

Figure 3. Simulated and experimental drain current as a function of the drain voltage for (A) p- and (B) n-FinFETs, for different fin widths, pre and post-radiation.

Some important figures-of-merit of analog performance for circuit applications should also be evaluated, such as transistor efficiency (gm/I_{DS}) and the intrinsic voltage gain (Av). Figure 4 shows that the n-FinFETs presents a degradation of the transistor efficiency after radiation, since the weak inversion region is inversely proportional to SS. For p-FinFETs after radiation, the opposite behavior of gm/I_{DS} was obtained since SS improves. Narrow fins of both types are almost not affected because of the immunity to the buried oxide charges already explained. At the strong inversion region, there is only a slightly variation of the transistor efficiency due to mobility degradation.

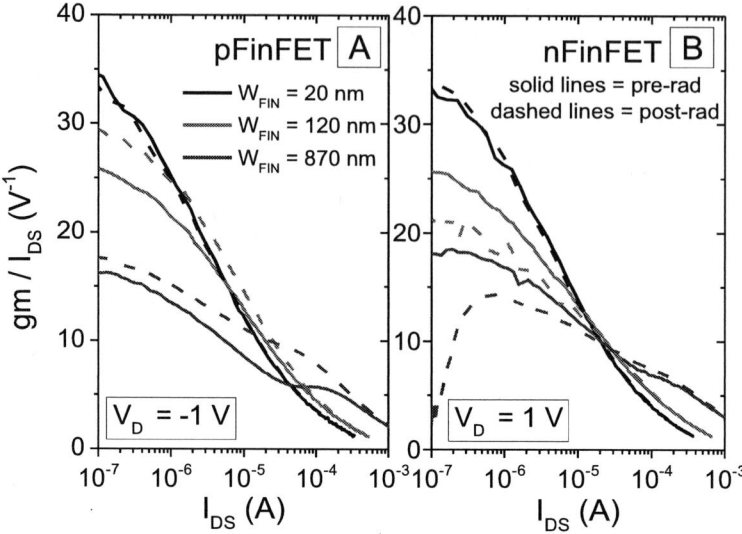

Figure 4. Experimental transistor efficiency (gm / I$_{DS}$) as a function of the drain current for (A) p- and (B) n-FinFETs, for different fin widths, pre and post-radiation.

The intrinsic voltage gain calculation is made following the mathematical relation shown in equation [1], where the gain is the result of the product of the efficiency (gm/I$_{DS}$) by the Early voltage value.

$$|A_V| = gm / I_{DS} \times V_{EA} \qquad [1]$$

The Early voltage values for each transistor were extracted at a gate voltage overdrive ($|V_{GT}|$) of 250mV and drain voltage (V_D) of 1V, and the obtained results are shown in Table I. For both types of transistors, narrow devices presented an improved V_{EA} due to the better coupling between gates, as expected. Comparing devices before and after radiation, the p-FinFET presented better values and n-FinFET presented worse values, following the behavioral tendency already explained previously.

TABLE I. Early voltage (V_{EA}) for p- and n-FinFETs for different fin widths, pre and post-radiation.

W$_{FIN}$ (nm)	p-FinFET		n-FinFET	
	Pre-radiation	Post-radiation	Pre-radiation	Post-radiation
20	20.21	22.10	10.52	7.07
120	3.44	3.99	3.98	3.70
870	1.53	1.64	1.75	1.75

From figure 5, it is possible to observe that the irradiation causes an improvement of intrinsic voltage gain for p-FinFETs and a A$_V$ degradation for n-FinFETs, for all inversion conditions, following the gm/I$_{DS}$ tendency. It is also expected that narrow devices present higher gain due to the higher Early voltage (V_{EA}), for both types of transistor.

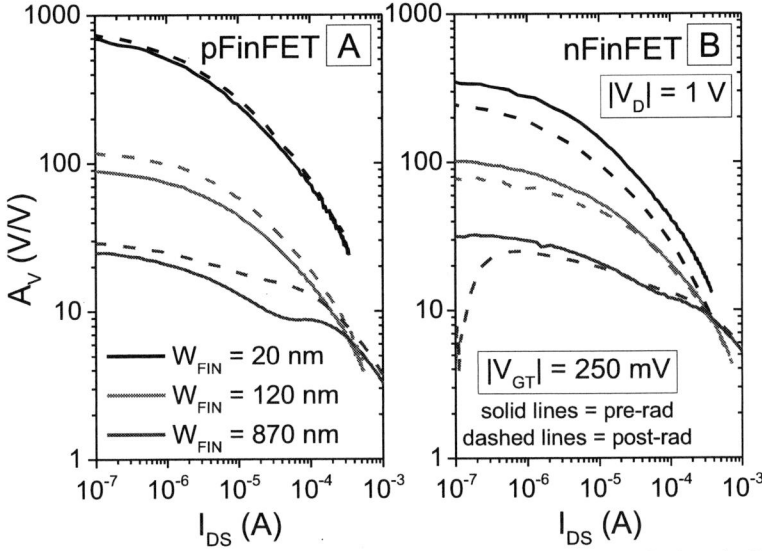

Figure 5. Intrinsic voltage gain (A_V) as a function of the drain current for $|V_{GT}|$ of 250mV for (A) p- and (B) n-FinFETs, for different fin widths, pre and post-radiation.

The evaluated current mirror circuit regarding the behavior with different loads (V_{LOAD}) is represented in Figure 6.

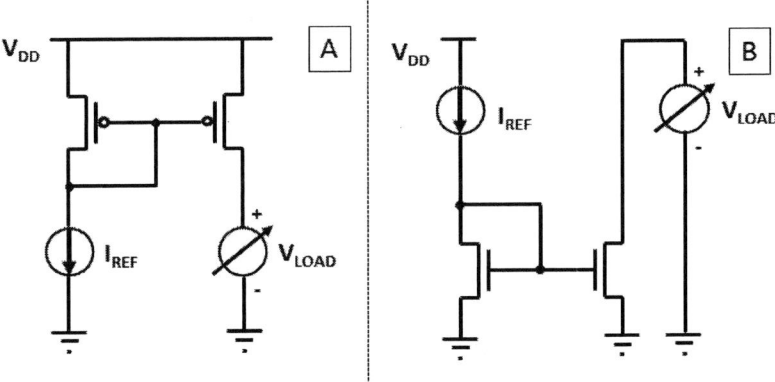

Figure 6. Testbench used for (A) p- and (B) n-FinFETs current mirror characterization in presence of different loads (V_{LOAD}).

Figure 7 presents the relation between drain currents (I_{LOAD} / I_{REF}), further referred as normalized drain current or drain current ratio, as a function of the V_{LOAD}, before and after the circuit has being submitted to proton irradiation. For this analysis, first the I_{REF} current of each transistor is defined for the perfect matching situation, i.e. drain current ratio of 1, in order to obtain $|V_{GS} = 1V|$ for both devices types and fin widths, before irradiation. From the insets of figure 7, it is possible to visualize that after irradiation, for the same applied I_{REF}, the perfect matching situation presents a slight variation in the normalized drain current and it is obtained for a more negative V_{GS} bias, for both types, due to the threshold voltage (V_T) change, promoted by the V_{th2} reduction.

In addition, from figure 7, it can also be noticed the influence of the change in V_{EA} on V_{LOAD} at the operation region of the current mirror. Since the V_{EA} decreases with the increase of the fin width, wider devices present higher current variations at the operation region. Even considering this effect, the maximum current variation obtained for the widest device is smaller than 10 percent, even for irradiated current mirrors. Beside the matching situation, it is also possible to notice that narrow transistors present a higher compliance voltage even for irradiated devices due to the better coupling between gates and consequently the immunity to the buried oxide charges.

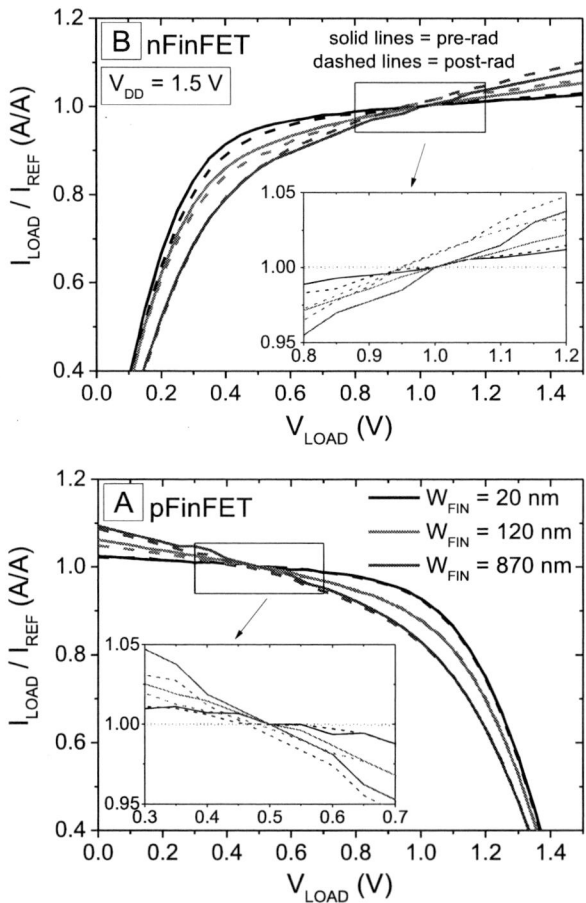

Figure 7. Normalized drain currents (I_{LOAD}/I_{REF}) for (A) p- and (B) n-FinFETs current mirror with different loads (V_{LOAD}), for different fin widths, pre and post-radiation.

Conclusions

The proposal of this work is to use a new approach for the simulation of circuits using FinFET devices based on the lookup table model using Verilog-A. The process consisted of a characterization of devices and construction of a table to be accessed by the simulation, allowing the simulation program to access the table and predicts the behavior of components using truly experimental data.

First, to affirm that this method is reliable, a comparison between measured data and the simulated data obtained was performed. Since both curves fitted with a great agreement, this method has been proven to be reliable. Then, analyzing the transistor efficiency (gm/I_{DS}) and the intrinsic voltage gain (Av), regarding the FinFET behavior for the different fin widths and the radiation matter, it is possible to certify that, based on previous works, both data have proven to be consistent as expected. Second, when it is analyzed the analog performance of the same devices within a current mirror circuit, based initially on the perfect matching situation before irradiation, the obtained result of losing the perfect matching situation after the irradiation because of the change in the V_T is consistent with the expected behavior. Even considering the radiation effect and the influence of V_{EA} on V_{LOAD} at the operation region, the maximum current variation is smaller than 10 percent, for the widest device.

Therefore, it is possible to conclude that the experimental data obtained are reliable and could be used to simulate the component behavior within a complex circuit. Further studies intend to demonstrate the application of the same method to create and analyze other circuits using FinFETs.

Acknowledgment

The authors would like to thank CNPq, CAPES and FAPESP for the financial support, and imec/Belgium for supplying the studied devices.

References

1. N. Paydavosi, S. Venugopalan, Y. S. Chauhan, J. P. Duarte, S. Jandhyala, A. M. Niknejad, C. C. Hu, *IEEE Access*, **1**, 201–215 (2013).
2. A. Korobkov, A. Agarwal, and S. Venkateswaran, *IEEE Trans. Comput. Des. Integr. Circuits Syst.*, **34**, 1696–1699 (2015).
3. D. Tassi, I. Messaris, N. Fasarakis, A. Tsormpatzoglou, S. Nikolaidis, C. Dimitriadis, *21st IEEE Int. Conf. Electron. Circuits Syst. ICECS 2014*, 710–713 (2014).
4. K. S. Kundert and O. Zinke, *The Designer's Guide to Verilog-AMS, June 2004*, Kluwer Academic Publishers, Boston (2004).
5. J.-P. Colinge, *FinFETs and Other Multi-Gate Transistors*, Springer US, Boston (2008).
6. J.-P. Colinge, *Silicon-on-Insulator Technology: Materials to VLSI*, Springer US, Boston (2004).
7. C. Claeys and E. Simoen, *Radiation Effects in Advanced Semiconductor Materials and Devices*, Springer GER, Berlin, Heidelberg (2002).
8. P. G. D. Agopian, J. A. Martino, D. Kobayashi, E. Simoen, and C. Claeys, *IEEE Trans. Nucl. Sci.*, **59**, 707–713 (2012).
9. L. F. V. Caparroz, J. A. Martino, E. Simoen, C. Claeys, and P. G. D. Agopian, *SBMicro 2016*, p. 3–6 (2016).
10. F. El Mamouni, E. X. Zhang, R. D. Schrimpf, D. M. Fleetwood, R. A. Reed, S. Cristoloveanu, W. Xiong, in *IEEE Trans. on Nuclear Science*, p. 3250–3255 (2009).

Author's Index

Agopian, P. G. D.	53, 65, 109, 121, 165, 171
Alcotte, R.	27
Alian, A.	27
Baryshnikova, M.	27
Bender, H.	59
Besnard, G.	129
Brunet, L.	129
Carmo, G. J. D.	53
Chang, T. F. M.	91
Chasin, A.	3, 59
Claeys, C.	45
Collaert, N.	27, 53
Cretu, B.	45
De Jaeger, B.	27
De Keersgieter, A.	3
de Sousa, B. R.	171
Dinh, T. V.	15
ElKashlan, R.	27
Ender, J.	159
Eneman, G.	3
Favia, P.	59
Fiorentini, S.	159
Fleetwood, D.	27
Fonseca, W. D. S.	121
Galembeck, E. H. S.	71
Gaudin, G.	129
Ghorbel, A.	129
Gimenez, S. P.	71
Goes, W.	159
Gomeniuk, Y. V.	151
Gonçalez, W.	165
Gudymenko, O. Y.	151
Gupta, M.	39

Hellings, G.	45
Hikavyy, A. Y.	59
Horiguchi, N.	3
Huynh-Bao, T.	3
Ingels, M.	27
Ishihara, N.	91
Ito, H.	91
Jang, D.	3
Khaled, A.	27
Khyzhun, O. Y.	151
Kisyl, D.	151
Korchovyy, A.	151
Kranti, A.	39
Kunert, B.	27
Kuo, Y.	143
Lacerda de Orio, R.	159
Langer, R.	27
Lin, W. S.	143
Linten, D.	45
Liu, L.	143
Loo, R.	59
Loup, V.	129
Lytvyn, P.	151
Macambira, C. N.	109
Machida, K.	91
Maitrejean, S.	129
Maleville, C.	129
Mannaert, G.	27
Martino, J. A.	53, 65, 109, 115, 165, 171
Masu, K.	91
Matagne, P.	3, 59
Mazen, F.	129
Mertens, H.	3
Misra, D.	135
Miyake, Y.	91
Mols, Y.	27
Nafaa, B.	45
Nazarov, A. N.	151

Nazarova, T. M.	151
Nguyen, B. Y.	129
Nikolenko, A.	151
Ogata, T.	91
Okholin, P. N.	151
Parvais, B.	27
Peralagu, U.	27
Perina, W. F.	65
Putcha, V.	27
Rangel, R. C.	115
Rangel, R.	165
Reboh, S.	129
Rodriguez, R.	27
Rosseel, E.	59
Rusavsky, A. V.	151
Sasaki, K. R. A.	115
Schwarzenbach, W.	129
Selberherr, S.	159
Shigeyama, R.	91
Sibaja-Hernandez, A.	27
Silva, V. C. P.	65
Simoen, E.	3, 27, 45, 53, 59, 65
Slobodian, O. M.	151
Sone, M.	91
Stepanov, V.	151
Sverdlov, V.	159
Takao, H.	79
Vais, A.	27
Vancoille, E.	59
Vasin, A. V.	151
Veloso, A.	3, 59, 65
Waldron, N.	27
Walke, A.	27
Wambacq, P.	27
Witters, L.	27
Yadav, S.	27
Yamane, D.	91

Yatskiv, R.	151
Yojo, L. S.	115
Yu, H.	27
Zhao, E.	27
Zhao, M.	27